In loving dedication to my beautiful wife Gayle Uyuni Sanger who's support, love and sincerity uplifted my passion, fueled my inspiration and transcended my motivation.

I would not be where I am today if it was not for her. She has been my guiding star, shining the way for me to grow and succeed. I am forever blessed, grateful and thankful to have her as my wife for the rest of my life.

Thank you mi amor...

In loving dedication to my son Valentino Sanger who means the world to my wife and I.

Birth of Magnetic Interest

As every child in grade school already enjoys magnets, this didn't quite peak for me until I attended physical science for the fist time. It was classrooms with this subject that would teach the basic introductions of physics. The more I learned, the more I was inspired. My mind would always be in a fast pace, constantly thinking, and expanding my thoughts to unfamiliar territory. I love the thrill for scientific discovery…a true passion of mine.

When I was a child all I ever wanted was to be an inventor. Inspired by numerous things, I was greatly influenced in my beginning years of life by a television show called Jimmy Neutron-Boy Genius. The show highlighted intelligence and creative innovation. Where his inventions made him the hero that saves the day in each episode. This fueled my passion tremendously. The mere thought for creating never before seen objects, innovations, and machines gave me the chills. My mind was full of ideas and theories.

My parents were patient listening to every single idea and theory that would cross my mind. They suggested I write down my thoughts on my creations into a journal. Which just so happened to be a great incentive towards writing my first journal. Later on in my adult life my wife became the one who inspired me to pursue the writing of this book. She surprised me by giving me a few journals as gifts. The journals allowed me to keep track of my thoughts and ideas.

Inspiration can come from anywhere or anything: being out in nature, having a conversation with another individual or even from watching an educational documentary. If you have the heart and passion to question and wonder about everything, or if you find excitement to seek out solutions for any and all problems your mind encounters. You are bound to create, invent, and to discover with the pursuit to improve or revolutionize man kind for the better. " To discover something brand new is to be the first and original student to learn what has never been learned before."

IDea, 46 Magnetic Flight

As I was watching an episode of History Channels original television show (Ancient Aliens), I encountered a mystery that needed to be solved. The episode I watched had to do with modern day UFO encounters and the eye witness testimonies. One of these eyewitness testimonies detained the bizarre maneuverability that a certain UFO possessed. This bizarre nature would be true throughout most, if not all, UFO encounters and eyewitness testimonies.

A few interesting points that was stated in their testimonies was the fact that for one: the craft possessed no sound and for two: the craft did not emit any kind of exhaust trails that would indicate some sort of combustion or chemical reaction that would lead to combustion for acceleration. The craft possessed thrust, velocity and acceleration. This was the mystery that I wanted to solve.

At this time, my mind began to race and I focussed my entire mental capacity to solving this mystery. I came to a logical conclusion that magnetism had something to do with it. Once I deducted that magnetism was the solution to this mystery I began expanding this idea which brings me to IDea, 46 Magnetic Flight.

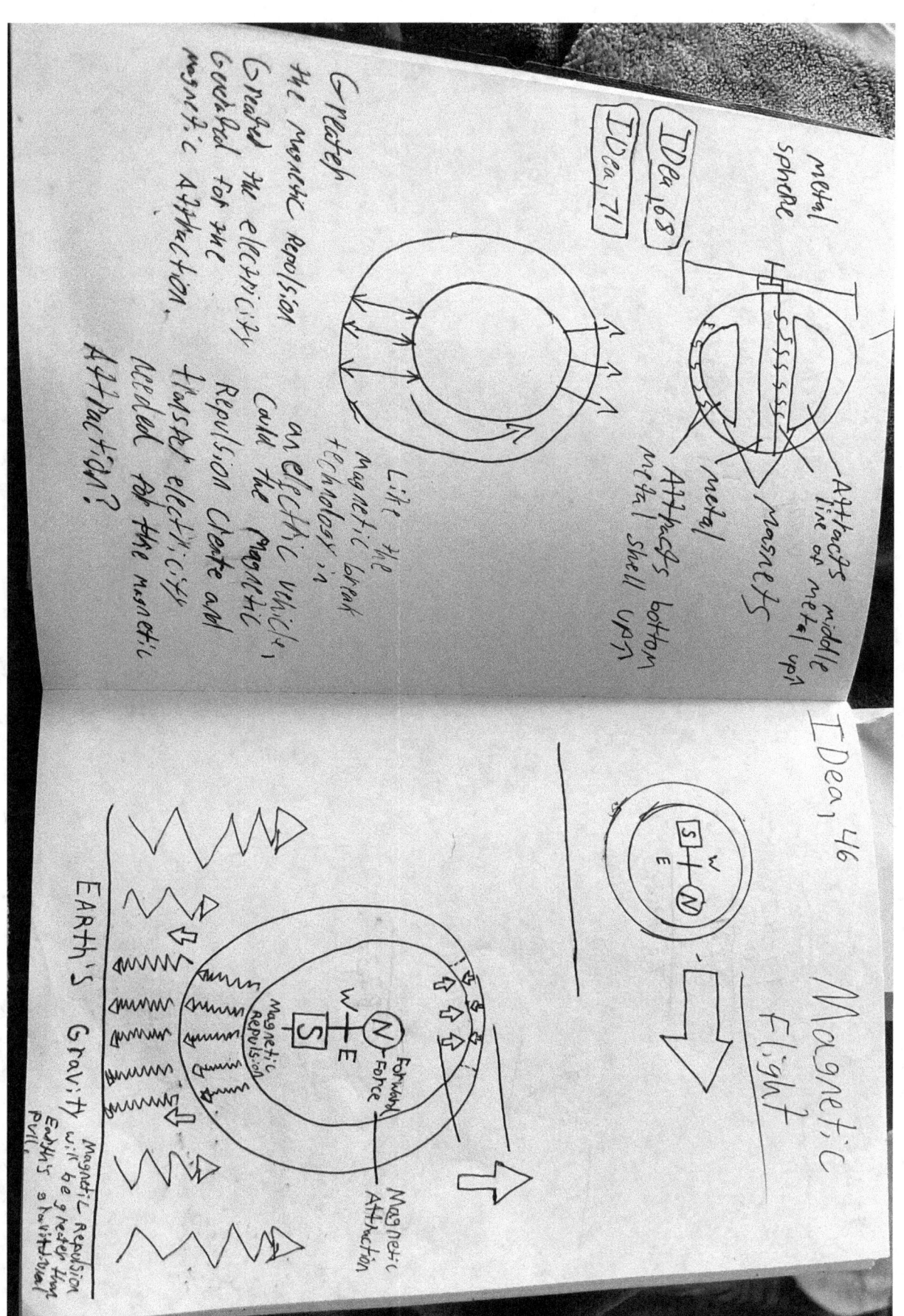

Idea, 68
Idea, 71

metal sphere — Attracts middle line of metal up
magnets
metal — Attracts bottom metal shell up

Greater the magnetic repulsion
Greater the electricity
Good idea for the magnetic attraction.

Greater the magnetic repulsion
Greater the electricity
Could the magnetic repulsion create an electric vehicle, thruster electricity needed for the magnetic attraction?

Like the magnetic break technology in an electric vehicle, could the magnetic repulsion create the thruster electricity needed for the magnetic attraction?

Idea, 46 — Magnetic Flight

Earth's Gravity — Magnetic Repulsion will be greater than Earth's gravitational pull.

Forward Force
Magnetic Repulsion
Magnetic Attraction

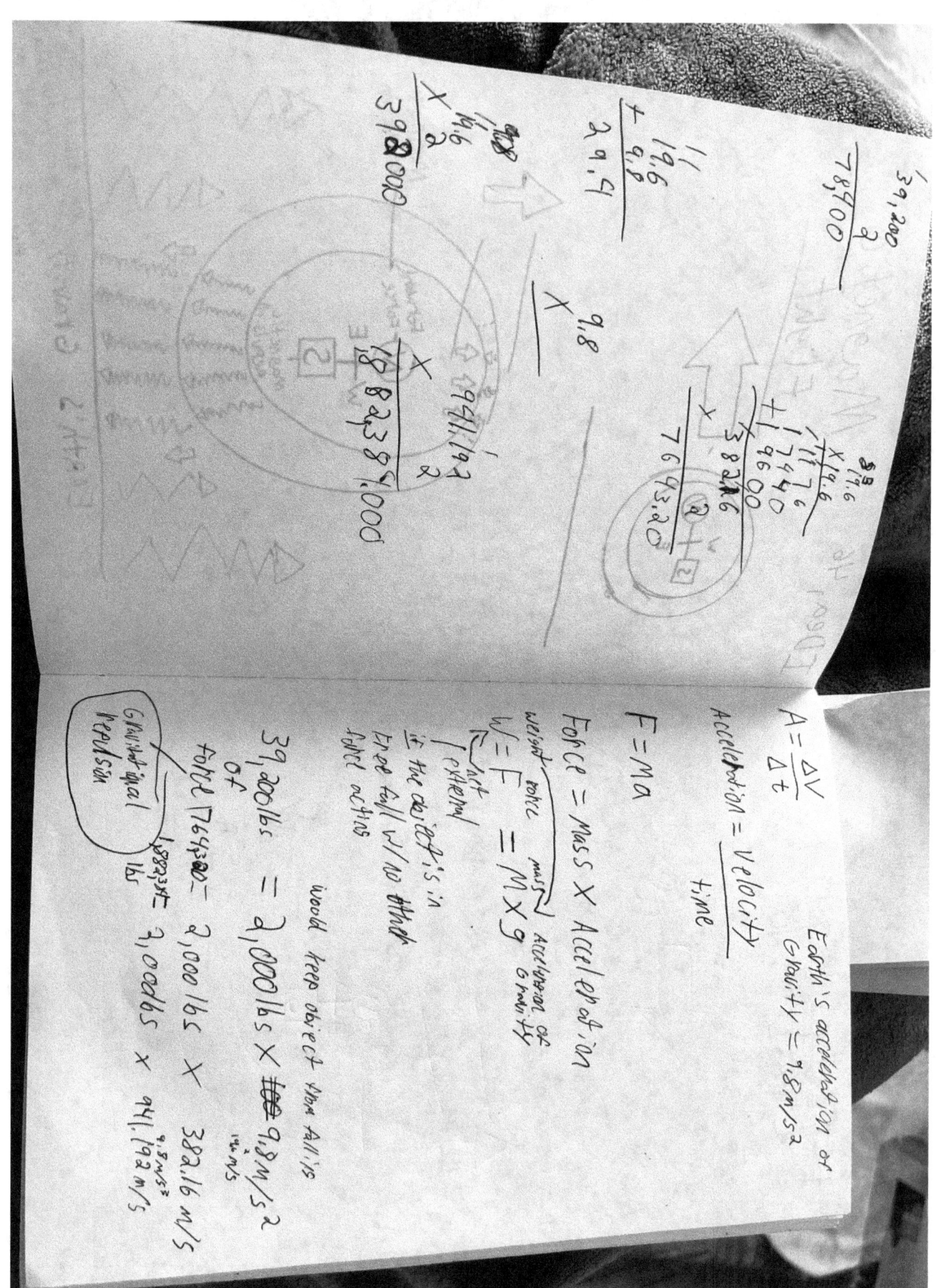

Earth's acceleration of
Gravity = 9.8 m/s²

Acceleration = velocity / time

$A = \frac{\Delta V}{\Delta t}$

$F = ma$

Force = Mass × Acceleration

weight   force   mass   acceleration of
$W = F = M \times g$ gravity
         [external
         if the object is in
         free fall w/ no other
         force acting

would keep object than fall is

39,200 lbs = 2,000 lbs × 9.8 m/s²
   of
Force 764,300 = 2,000 lbs × 382.16 m/s²
      1,882,395 = 9.8 m/s²
      lbs   × 94,192 m/s

Gravitational
revulsion

$$F = \left[\frac{B_0^2 A^2 (L^2 + R^2)}{\pi \mu_0 L^2}\right]\left[\frac{1}{x^2} + \frac{1}{(x^2 + 2L)^2} - \frac{2}{(x+L)^2}\right]$$

Force between 2 identical cylindrical bar magnets placed end to end

B_0 = flux density very close to each pole, in T.
A = area of each pole, in m^2.
L = length of each magnet, in m.
x = separation between 2 magnets, in m.

$B_0 = \frac{\mu_0}{a} M$ relates the flux density at the pole to the magnetization of the magnet

How much thrust (magnetic repulsion) is needed to break free from earth's gravitational pull?

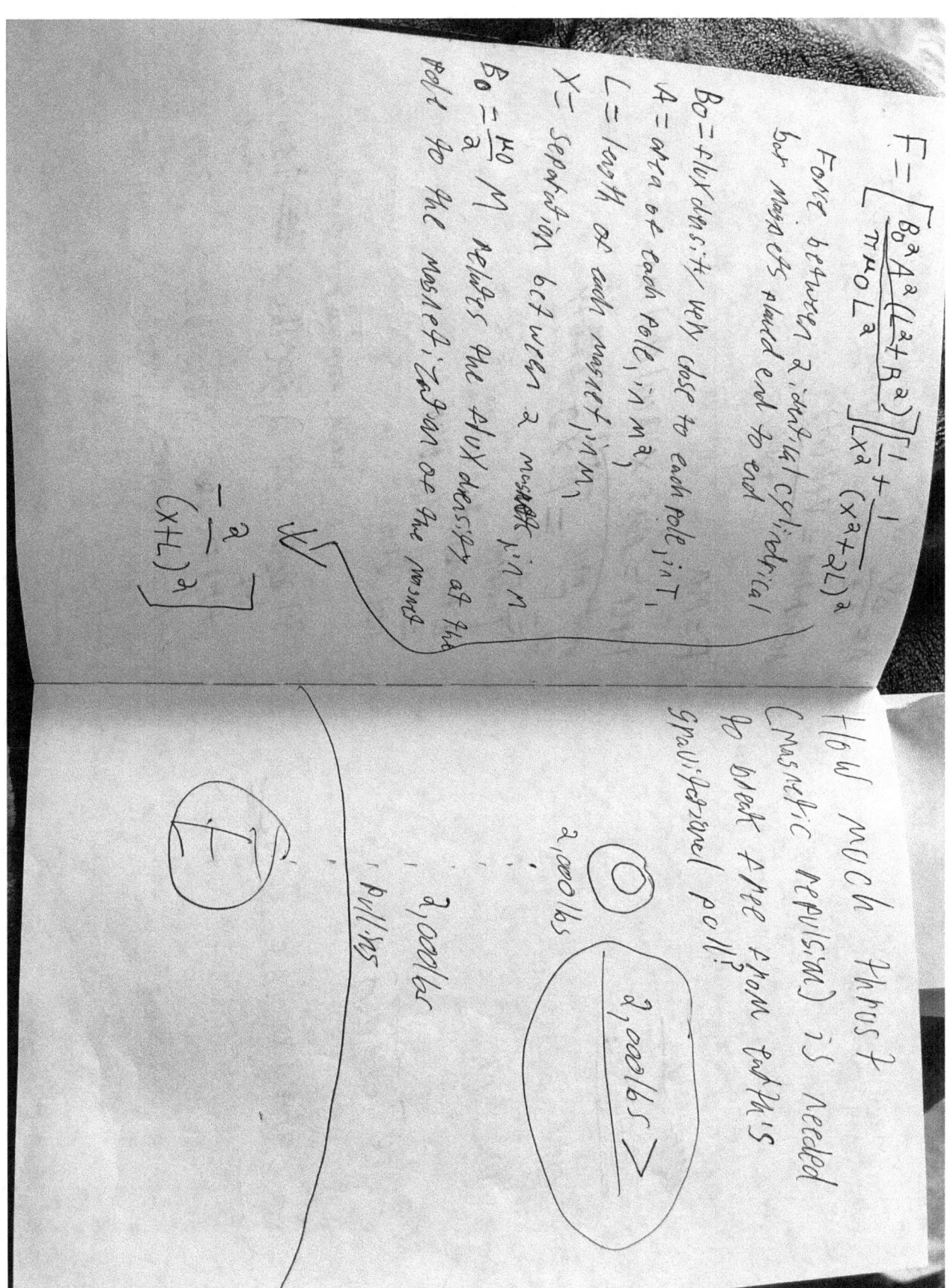

2,000 lbs
2,000 lbs
pull, lugs

$2,000 \text{ lbs} \times 16093 \text{ m/s} \over 3600 \text{ per hour}$

$89,407.7778$ Newtons of ~~salt~~ lbs

to move 2,000 lbs at 100 mph

$89,407.7778 \div 4.45$

$20,091.6355$ lbs of ~~~~ thrust

Application examples:
Bike: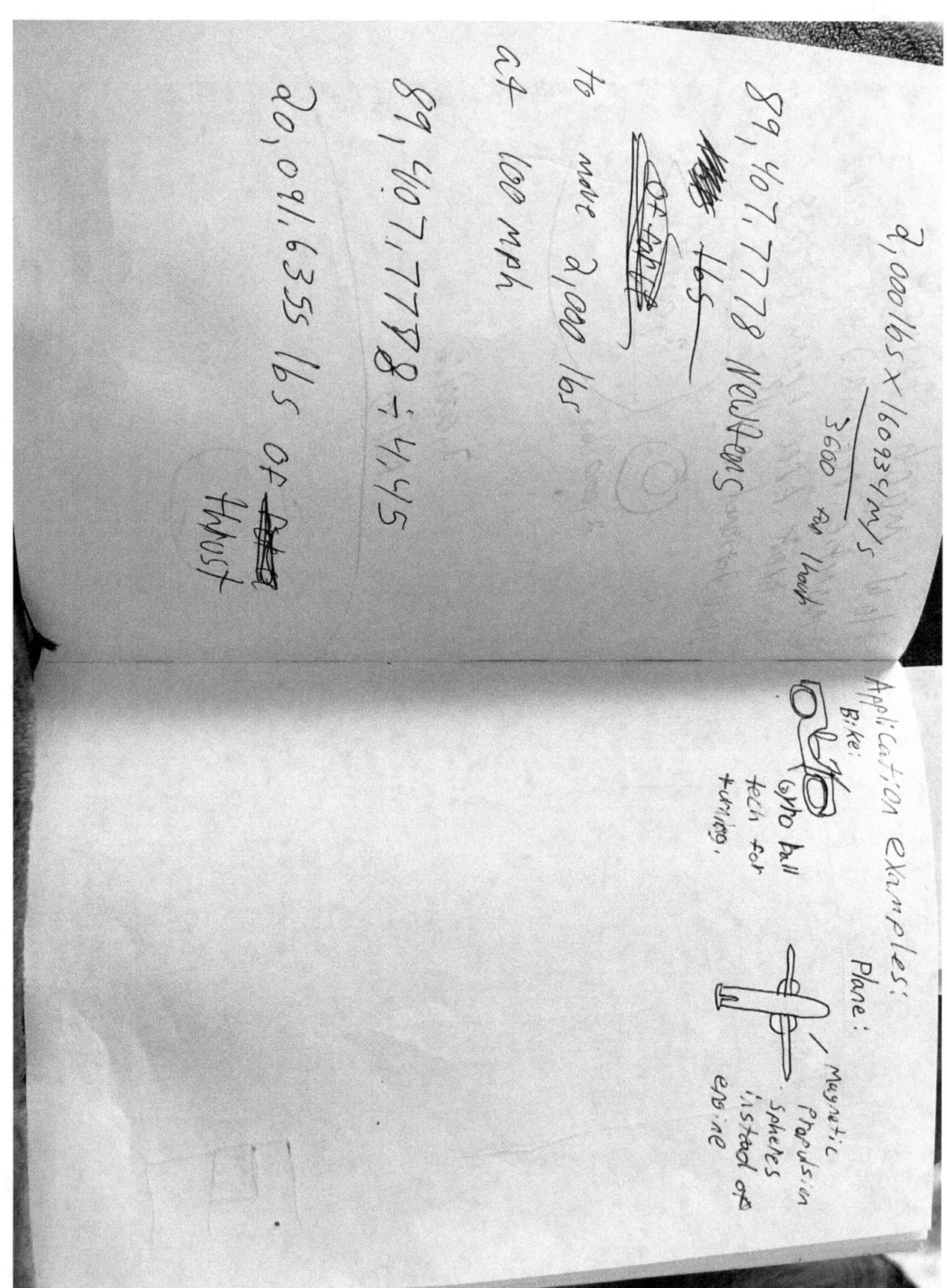
Gyro ball tech for turning.

Plane:
Magnetic Propulsion Spheres istead of engine

And that where I stopped. I was trying to solve the physical problems of this magnetic flight IDea. I tried not to be so hung up on the mathematics. I figured that if one day we "mankind" got to the point of a physical prototype, that only then would getting real time numbers for the equations would become more relevant. I gave it a shot anyways.

With the magnetic gyroball taken shape as the solution for Magnetic Flight, for now, I began applying this "innovation" to several fields.

Magnetic Flight Gyroball Applications

From cars and motorcycles to boats and aircrafts. The applications seemed endless. I wanted to keep my thoughts close to the ground of reality and practicality rather than high in the sky of sci-fi driven creations. To do this I must consider several types of regulations and of course the safety requirements if I wanted the conception of my ideas to become real and legitimate.

For example, programming the magnetic gyroball to limit the levitation to only a few feet off the ground, seems very simple right? This would allow certain highway and road driven vehicles to still obey the rules of the road. We must baby step the flying cars. I have observed certain regulatory requirements that have limited innovations today.

They have already built prototypes, or at least have created plan in blueprints, for remarkable vehicles that can take to the air with systems similar to that of helicopters. However, airspace regulations have shut them down, for now. For me the excitement is real and I can not wait for the day in which a company figures out a way around this necessary regulations.

When I first saw the computer generated prototype, about 7 to 8 years ago, I thought for sure the world would see flying cars within ten or fifteen years. A few points did raise that would put a halt, or at least temporary pause to this grand idea.this specific innovation however requires a lot of air space and shoulder space.

In the computer generated demonstration, the vehicle was first shown in an empty parking lot and then the wings opened up, this is the part that requires the shoulder space for take off, very similar to that of a helicopter. This is also true for the space needed to make a landing. In essence the innovation requires a helicopter pad or designated zones for arrival and departure.

Now, I am not shutting this idea down. I love science, technology and innovations as much as the next person. I do believe in man kind's ability to build, create and fund anything we want. If we as a society truly wanted to build, create and fund to accommodate to the challenging regulations which slows down progress, reasonably for safety purposes, we can.

Massive funding fueled by popular demand is just an example of salvation for these great ideas and innovations. If we developed the technology like the magnetic gyroball, that might be enough to get a substantial amount of individuals excited enough to bring to life that innovation. For now, let's start by showing them the possibilities of the near future.

Introduction to Architectural and Reusable Energy Applications

Aside from Magnetic Gyroball applications and the constant technological challenge, that is permits and safety regulations, I would like to skip ahead into some architectural and reusable energy applications: the few applications entail the following:

"IDea, 53 Magnetic Stairs", "IDea, 54 New Era of Construction", "IDea, 56 Hydro, Wind and Solar Turbine", "IDea, 57 Magnetic Alternator", "IDea, 58 Portable Electricity/Magnetic Generator", "IDea, 61 The Architectural Platform", and "IDea, 160 Reusable Energy Within Water Pipe".

IDea, 53 Magnetic Stairs

IDea, 53 Magnetic Stairs is a basic application for the "near future" of magnetism. Later on in this book I will uncover the magnetic technology that will make this idea a reality with "any" material. Allow me to translate for you what the schematic is and how it would work. Important thing to understand is, the material of the "Magnetic Stairs" varies. We all know the idea of floating "metal" stairs has been around all over science fiction, but with any material? Short answer, yes its possible.

Idea 53 Magnetic Stairs

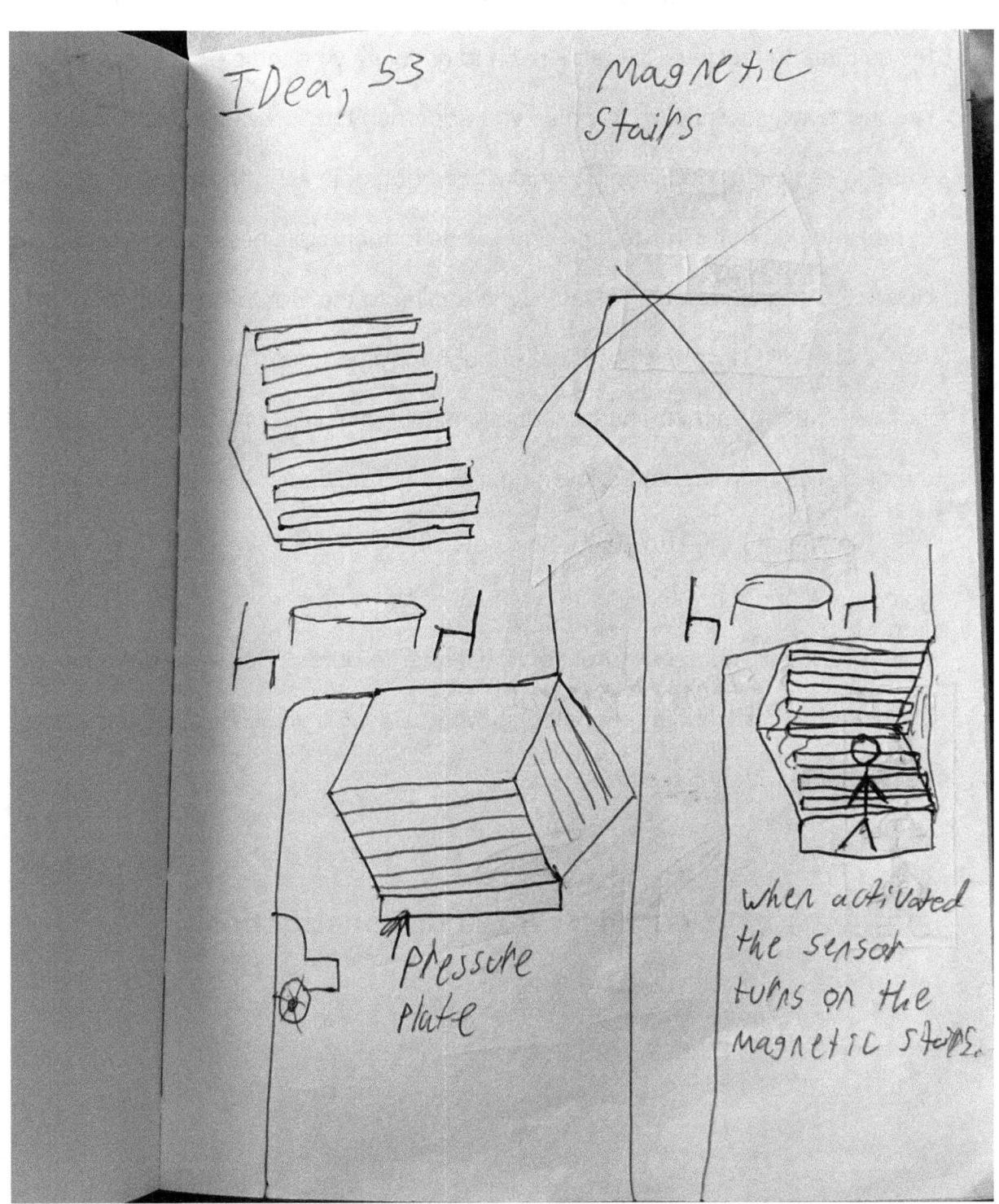

Pressure plate

when activated the sensor turns on the magnetic stairs.

Simply put, we have a two story building and the first appearances would tell us that the first level is clean, flat and absolutely no indicator of reaching the second level. However, as soon as you approach the "pressure plate" and or "motion sensor" an electronic signal is sent out, activating these magnetic stairs to rise into position. This technology will be further explained into the book. "IDea, 53 Magnetic Stairs" gives new meaning to the word minimalistic.

That is all I wrote down, just enough to remember. There is no science included with the schematic, no equations to explain the physics, nor is there mathematics detailing and explaining the constructive aspects of this project. With the way in which my mind operates, I occasionally refer to several ideas, going back and forth, to try and eventually paint a much bigger and brighter picture. You will see an example of how I refer to other "IDea's" to get a better understanding of certain technologies with "IDea, 54 New Era of Construction". Let us begin with the sketch.

IDea, 54 New Era of Construction

IDea, 54.
(IDea, 71)

Refer back to
IDea, 46.

Magnetic flight

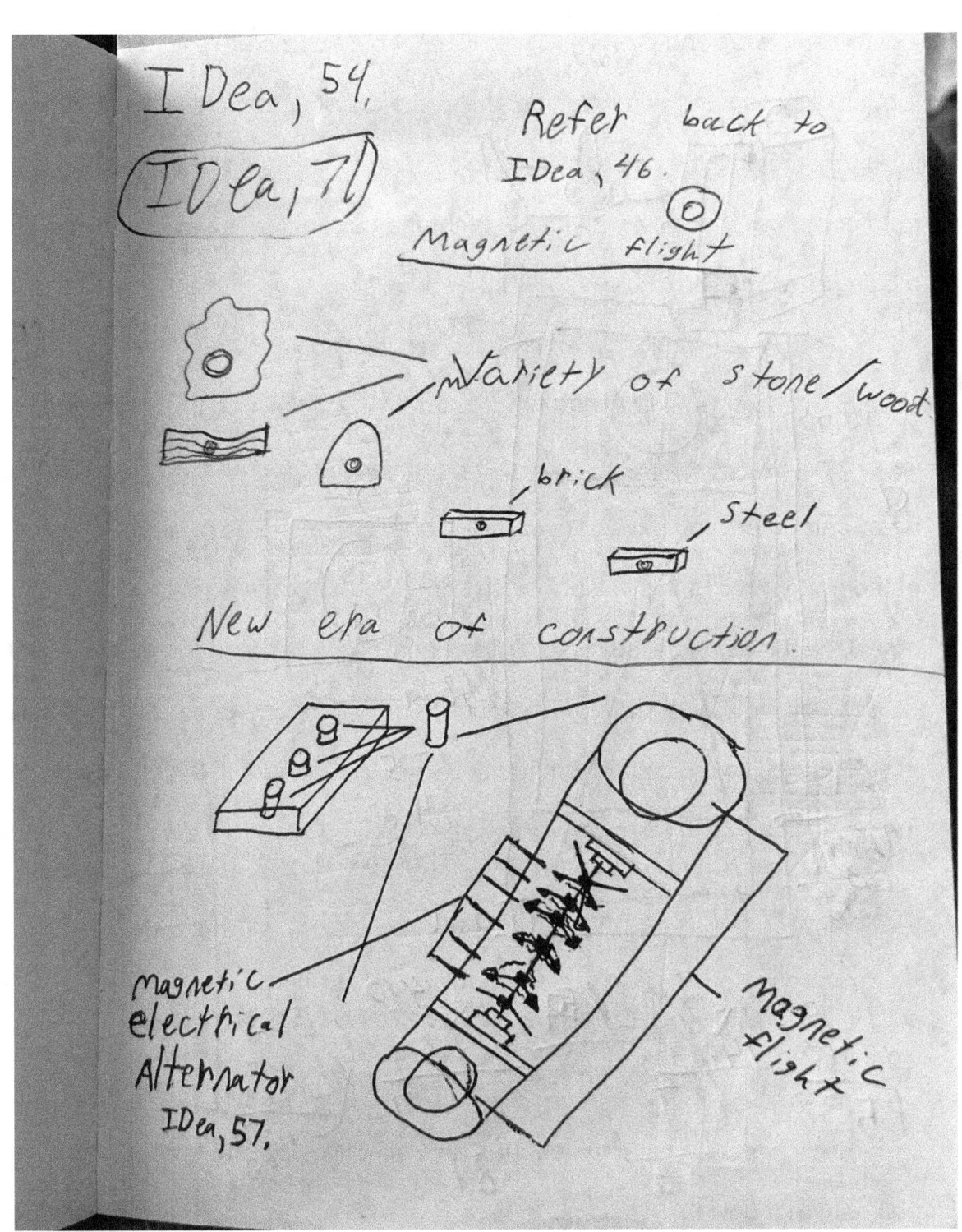

Variety of stone/wood
brick
steel

New era of construction.

magnetic
electrical
Alternator
IDea, 57.

magnetic flight

At this point my mind starts racing and my heart starts pounding with excitement, hope and promise. This "IDea" would completely revolutionize modern day construction. Imagine, the entire construction project would piece itself together like the pieces to a puzzle. Bring to life an architectural design, from computer to reality, with just a push of a button. Magnetic Flight would be the key to all of this.

The pieces would levitate themselves into position and depending on the building material would dictate the constructive specialist to finish the job. If you send out a few welders you have got yourself a steel foundation. If you send out a few carpenters you have got yourself a house. There are countless examples of the betterment of this 'IDea". As you continue reading you will realize that I organized this book based on the "IDea" number rather than the category of the "IDea". I cannot control what "IDea's" come into my mind or the order in which I think of them. I write them down in my journal in the order in which I generate them. Which is why I refer back and forth a lot. Important details to keep in mind as you continue reading.

IDea,56 Hydro, Wind and Solar Turbine

This next "IDea" brings about a new effective way at harnessing the power of ,not only the wind and the sun, but also harnessing the raw power of our oceans and the tidal forces. I have seen already ideas that take the hydro turbine under the water to capture this natural untaped resource. However, there is a problem with this idea which keeps it from launching. The challenge is our oceans wildlife and our oceans plant life. Any rotating turbine needs to be untouched or unblocked in order for it to be fully effective. The underwater hydro turbines can also pose a threat to the oceans wildlife. How can you harness such a great natural power while posing no threat to the ocean and its wildlife? I have a solution to this problem.

Idea, 56 — hydo, wind, and solar turbine

?— could put a regular wind turbine on top to get the maximum amount of energy.

— the mechanism floats on the surface.

Moves up or down with the change of water elevation / change of tide / change of sea level

— the paddles rotate which turn physical rotation into electricity

electrical line underground

As you can see ,within my sketch, there are no moving blades or turbines below the surface of the water. The paddles that rotate only penetrate the surface of the ocean, just enough to harness its immense energy. This system floats on the surface of the ocean and moves up and down according to the tides and or the waves, thus maintaining maximum efficiency of electrical energy to be generated. I could not help but include the possibility of a wind turbine and a solar panel on the top of this innovative structure. My next "IDea" brings about a new way of generating electrical energy that does not require water, wind or solar energy. This is a magnetic "IDea" that represents a brick on this long road of magnetic possibilities. As you continue reading, you will be taken down that path of innovative and revolutionary technologies. It is important, I must add, that these simpler "IDea's" are brought up in this book to show to you how my mind evolved these thoughts and "IDea's" into something that can revolutionize mankind, if my theory is correct. You will learn and understand more of what I am telling you later on in this book.

IDea, 57 Magnetic Alternator & IDea, 58 Portable Electricity/ Magnetic Generator

Idea, 57.   Magnetic Alternator

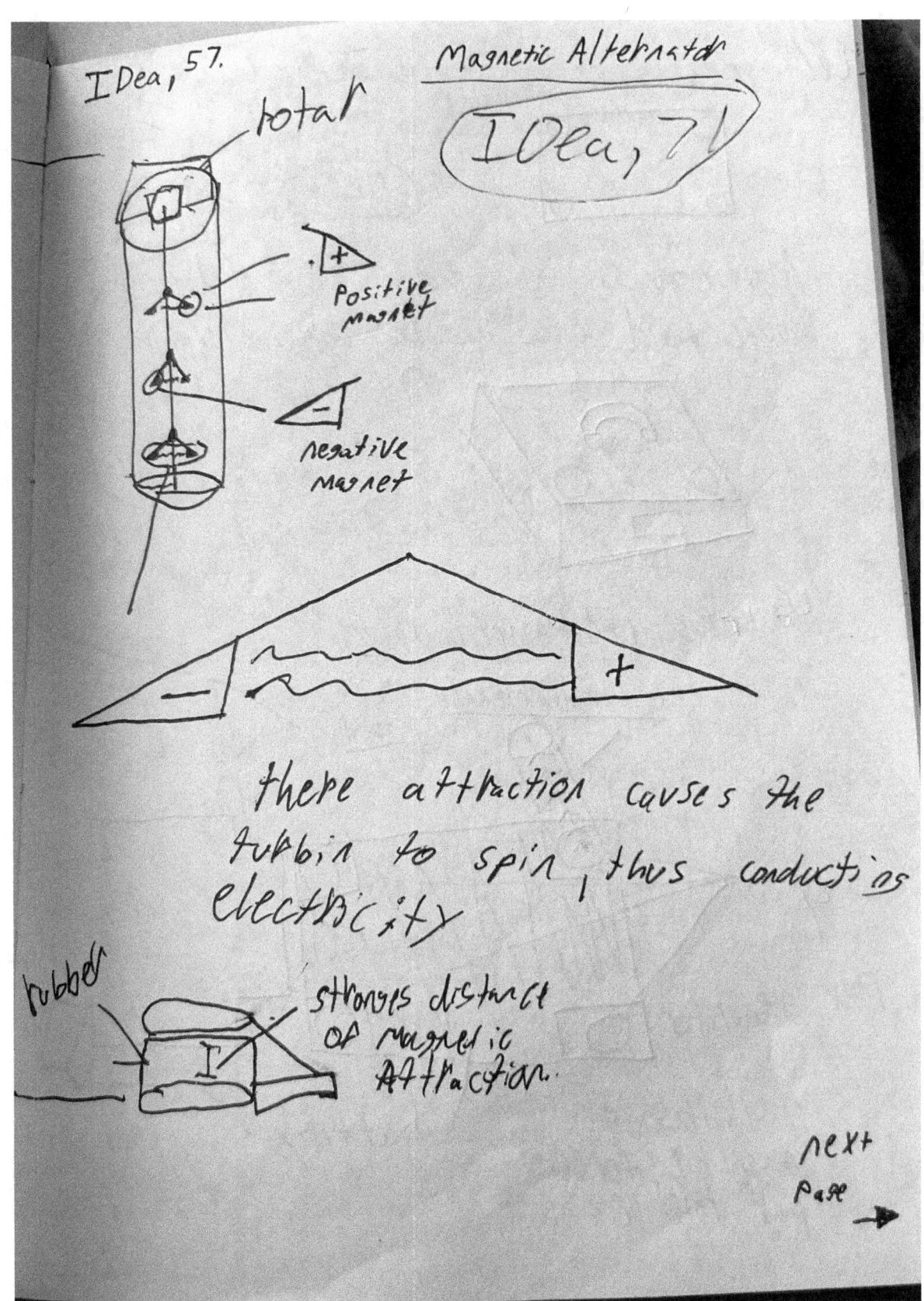

Idea, 7!

there attraction causes the turbin to spin, thus conducting electricity

rubber — stronges distance of magnetic Attraction.

next page →

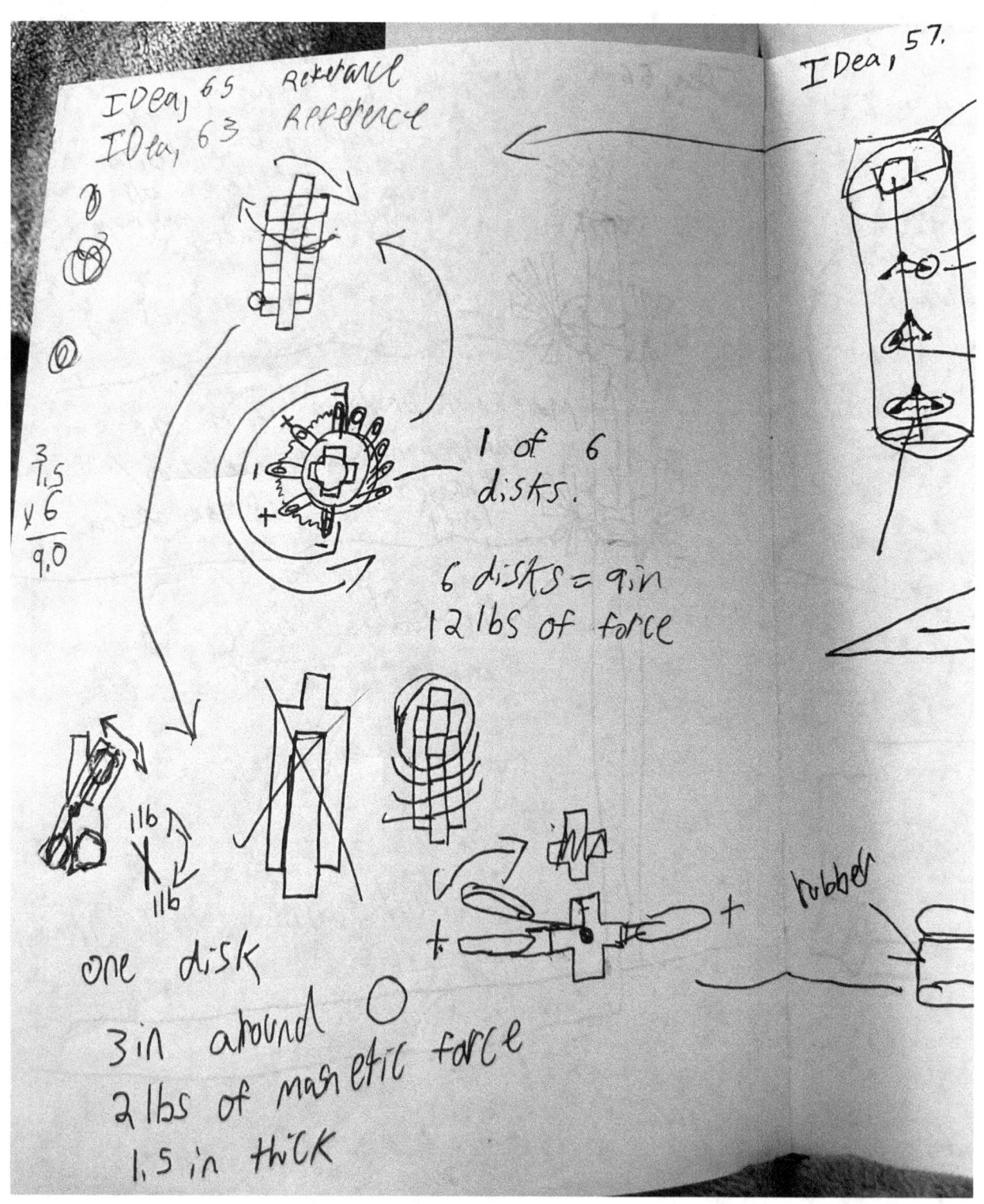

This time I joined together two "IDea's" simply because they are the first example of "hand to hand" applications. The concept of "IDea, 57" is that using physical rotation through magnetic attraction could give us electrical energy wherever we are regardless of weather conditions, I would elaborate in more detail on a similar project that would be "IDea, 63".

As soon as I theorized one technological development, my thoughts began to race. What would having this "innovation" mean for mankind? "IDea, 57" would grant "free" electricity in every corner of the planet. Modern day reusable energies require sunlight, the wind or even ocean currents to generate electricity. What about areas on Earth, or even Space, that do not have the natural elements necessary to generate electricity? My answer is unlimited physical rotation powered by Magnetism. For now my answers come in the form of a few simple sketches.

I want to develop on "IDea, 58" and the potential it has. Devices like "IDea, 58" would be convenient, easy to operate and easy to carry. There will be no need to buy gasoline for this generator. This "key" difference would make it, not only safer for the consumer, but also would make it more affordable. The only thing that a person would have to pay for is the occasional maintenance, through research and trials will give that aspect more longevity of use for the consumer and will keep more money in your wallet.

Idea, 58.

Free electricity
Magnetic rotor
Magnetic generator

Portable electricity

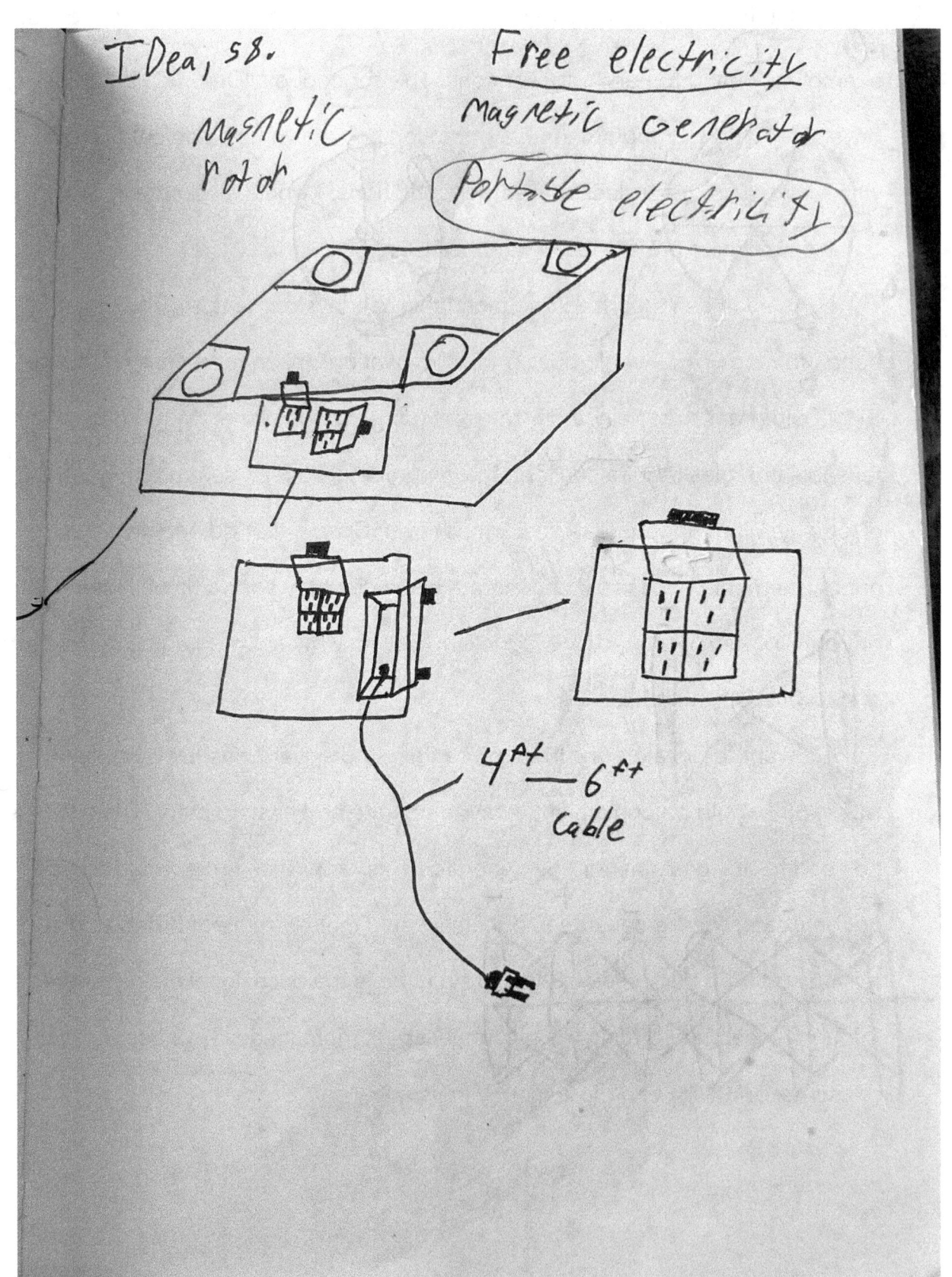

4ft — 6ft cable

Idea 58.

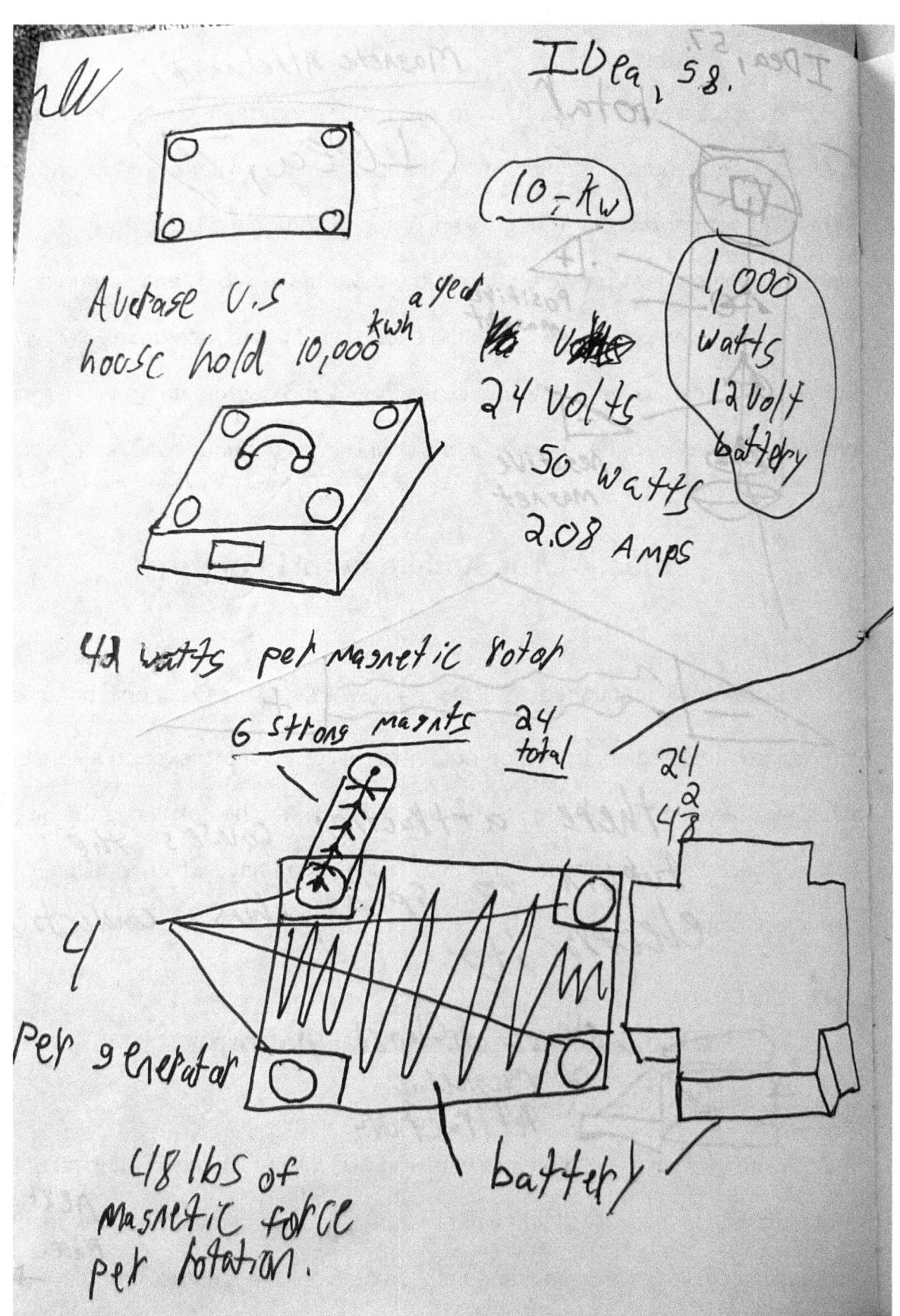

Average U.S household 10,000 kwh a year

10-Kw

24 Volts
50 watts
2.08 Amps

1,000 watts
12 Volt battery

42 watts per magnetic rotor

6 strong magnets — 24 total

4 per generator

48 lbs of magnetic force per rotation.

battery

24
4⟌2⁸

Now, as a reader, I want you to keep your mind open. Grasp the "IDea's" I give you throughout this book and try to implement your very own applications. Please do not narrow your thoughts and imagination on the basic and elementary sketches I provide. For me, my thoughts are "big" and "complex". There is simply no possible way I could ever draw, in complete detail, my "vivid" imagination. I can verbally develop it, describe it and coordinate how it can be created. Please keep these words in mind when you continue reading this book.

IDea, 61 The Architectural Platform

I previously mentioned, in "IDea, 54 New Era of Construction", bringing to life Architectural design, from a computer to reality. I want to expand a little bit of what that "Architectural Design Software/ Platform" might look like in the near future. This comes in the form of another "IDea" that I have which is "IDea, 61 The Architectural Platform".

IDea, 61 The Architectural Platform

Inspiration: Halo 3 Forge, Minecraft, Sims

A Software that when building a structure in 3D, will tell you how much all of the materials will cost. It will tell you what materials are being used and how many materials are used. It will simultaneously create your blueprint to your 3D structure. If possible, tell you how many labor hours it will take to construct in a certain time frame. Large companies can send a project design request with a price range guideline and key notes from the company for what they need and are looking for. That request will appear as a notification on the top of the screen of all the "creators" t.v. you can switch from "creator" to "locator" mode to browse other designs by other people. This would be an interactive interphase. If a company likes your design, a business deal can take place. 10% of the purchased price from the deal would go to the software/ platforms company for creating the software/ platform, a third party sales fee. The software/ platform will be available to everyone. The software will use real world physics to ensure the strength, the durability and practical functions of your design. From corporate buildings and offices to homes and recreations. Those designs can be purchased for real world construction. The designs can also be sold to be used in films, animations, and/or video games. The uses for the designs would carry a long list of possible buyers for numerous projects.

    if this "IDea" became a reality it will ignite the passion for all creative freelance designers that cannot afford college level schooling. It will open up many doors and possibilities for creative people to earn more income while continuing there own jobs and/ or careers. This "IDea" might just be the answer and salvation for those that cannot find work or for those that become unemployed. Please let us join together, revolutionize modern society and create the "New Era of Construction".

"Need not to have a degree... Need not to be a qualified Architect... All you need is your imagination, creativity and passion to design... To be successful"
-Zachary Sanger

IDea, 160 Reusable Energy Within Water Pipe

This was a recent "IDea" that took place after my rough draft of this book. I wanted to include this "IDea" because I realized it was too important to not be mentioned. This is by far one of the simplest reusable energy "IDea's" I have, but it is far more affordable than the others and can have the biggest impact. The schematic is as follows.

Idea, 160 — Hydro turbines within the pipes of ~~our~~ homes, businesses, or facilities. Goal? to generate electricity for yourself and of the grid.

Locations? Primary location would be directly connected ~~to~~ from the water tank to the rest of the line. (area of most force)
Secondary locations would be faucets, ~~the~~ toilets, showers, dishwashers.

"reusable energy within water pipe."

illustrations continued ⟶

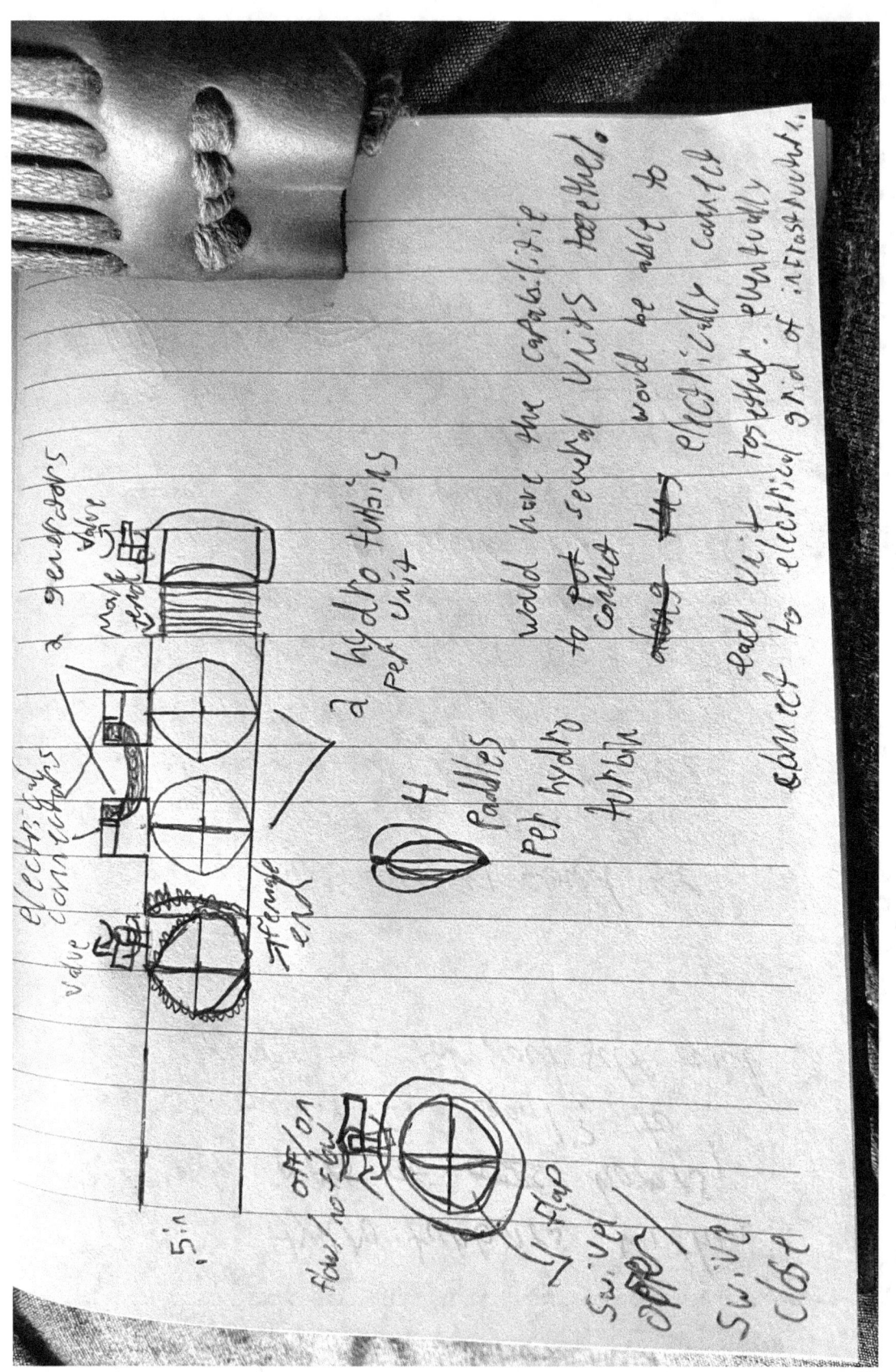

would have the capabilit[y] to put several units together. Would be able to connect them electrically, coupled together. Eventually each unit together. Eventually connect to electrical grid of infrastructure.

This reusable energy product is the key to get everyone, of any income, involved with the "green" movement. Something that is very special about this "IDea" is that it will not only be affordable but the amount of electrical power that will be produced will be substantial. Imagine if every household in your town, city, state, or country had this product installed. It would be the worlds largest "Hydro Electric Water Damn" that would not come at the cost of destroying parts of the land nor potentially having dire environmental drawbacks/ devastation. In a way, this "IDea" is non-bipartisan and would be accepted by everyone regardless of political/ moral preferences. Speculatively, the vast majority of individuals love the idea of owning solar panels and/or wind turbines. However, these routes are very expensive which makes it "exclusive" rather than "inclusive". This "IDea" will make the "Green Movement" more equitable. I must add that it takes longer to generate a financial return towards these expensive investments. For one, the high cost of equipment, installment, and to maintain over the years. And secondly, the power generation depends on "perfect" weather conditions.

My "IDea" is inclusive and it highlights community effort to generate massive amounts of electrical energy. For the states that participate, the people will no longer have to pay for electricity. In fact, the states may even install them for you free of charge, if it means vast communities generating massive amounts of electricity. Understand, to build a Hydro Electric Water Damn, costs billions of dollars, disturbs/destroys the environment, and there's naturally an available land/water limit to build several of these damns. Knowing this, governments/electric companies across the world, may give this to you for free. If not, it will be very affordable and highly electrically generative compared to its expensive counter parts (solar panels and wind turbines).

Perfect This... Perfect Space Travel
IDea, 59 Magnetic Gun

Now, I made a personal note that was not mentioned at the time I drew this "IDea". I stated "perfect this... perfect space travel", I do not want to disengage anyone with this "near cited" weaponized application. For the functions will also operate for space travel. To use as a way to magnetically launch and/or propel Space shuttles. From weaponized projectiles, to advance our efficiency and effectiveness against global terror threats. To a ship in Outer Space utterly to be carried out to vast distances, at incredible speeds, without the use or "need" of complex and costly fuels.

In my personal opinion, combustion "will" be obsolete for space travel. Strictly because combustion will require constant surplus when venturing far out in Space and far away from Earth. Reusable energy and Magnetic Propulsion is the key for space travel. Before I "drift" off topic, let us get back to IDea, 59

Magnetic Gun. I want to demonstrate through illustration the magnetic functions that would also be applied to space travel.

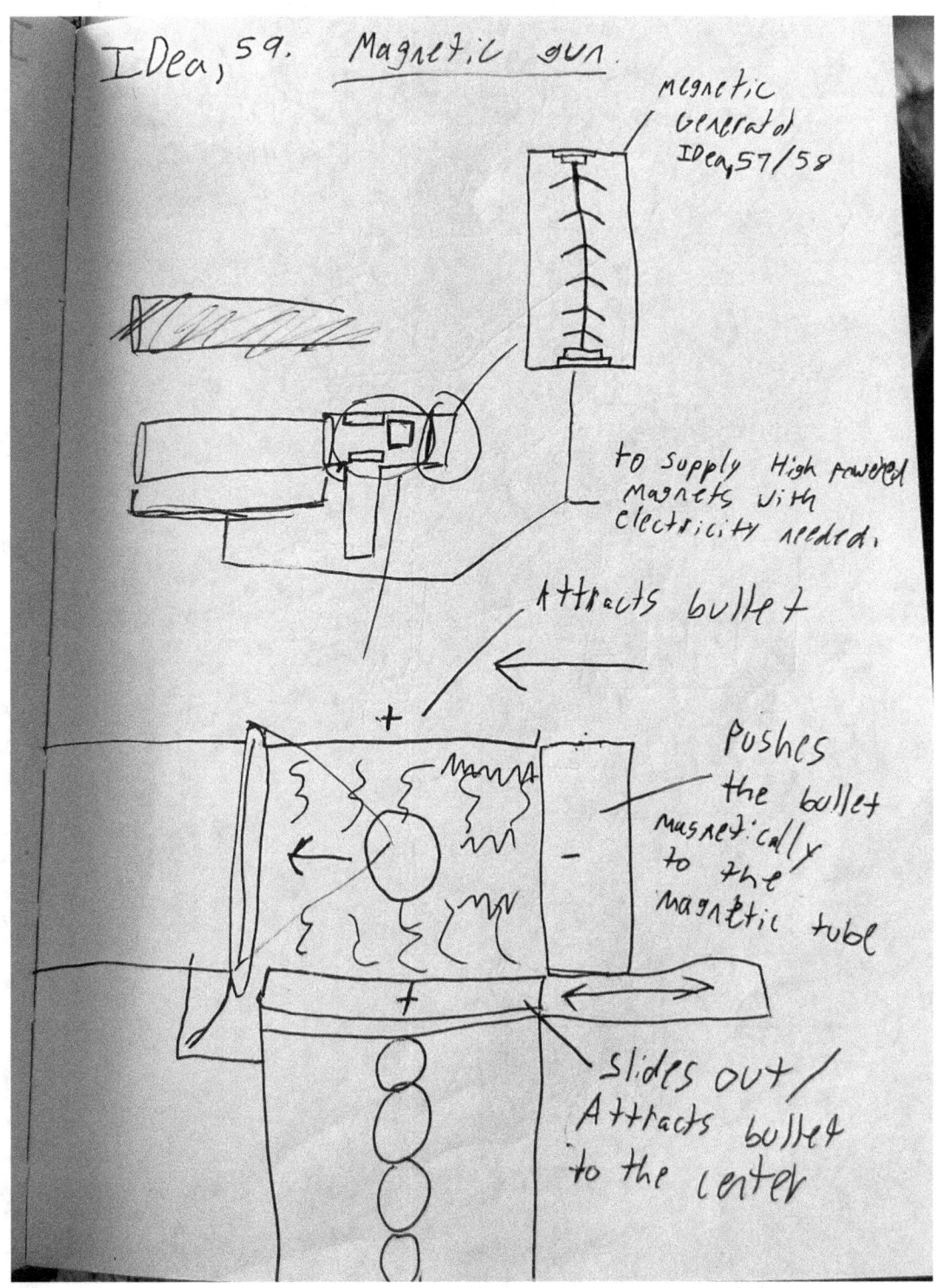

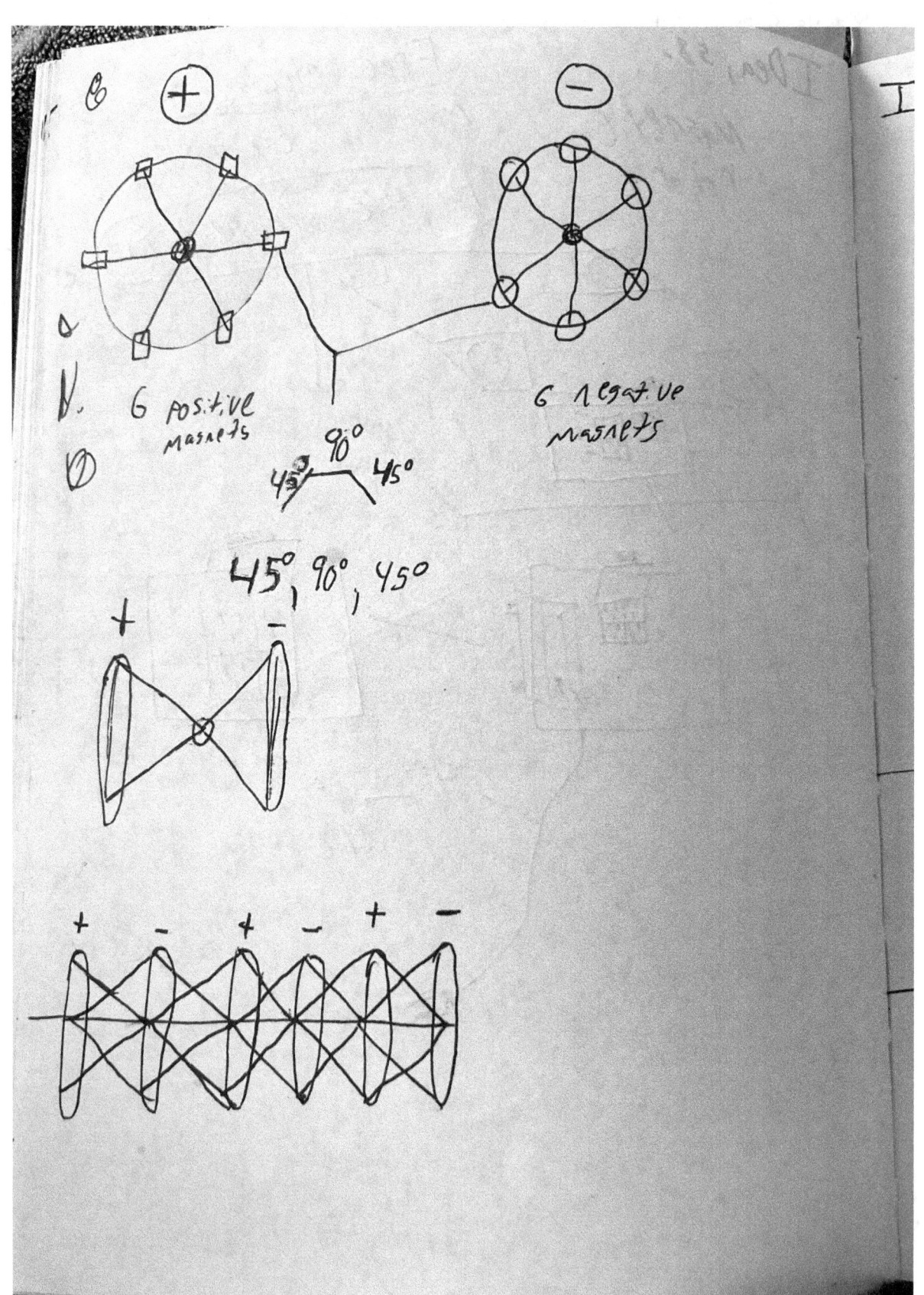

If you enlarge many aspects of this sketch 100 times, or even 500 times, you will have a size worthy magnetic infrastructure to launch space crafts to greater and newer distances. This method will be highly efficient and overtime will save lots of money on "combustion" resulting fuels and the expensive "extraction" of such fuels. Additionally, we can produce these magnetic infrastructures for the return trip back to Earth. In essence, we would be creating "train tracks and train stations" in Space that would complement our "commercial" destiny in Space.

You can apply such devices and technology to combat threats, both natural and intellectual, to our planet. With enough velocity and a strong enough payload, we can protect our planet at an impressively "safe" distance away from any destructive debris that could have major repercussions back on Earth. This will highlight mankind's eagerness for efficiency and getting the job done right the first time, measure twice and cut once.

Now, let's shrink the sketch back down. The original "IDea" was to be put into the hands of a soldier, a military application, "the revolution of the gun". Of all of the benefits this magnetic gun would have, the key benefit to me was the "true" silence each shot would have. Silencers now a days can be bulky and give unnecessary weight on the gun bearer. This is an aspect that can certainly be improved on.

A few more benefits that stood out to me was longevity and health of the firearm, via no "physical" mechanisms that would "heat" up or jam. This problem can delay not only efficiency but also delay the dire need of constant firing to push back military opposition. Also, there will be no need to clean your equipment as vigorously as you would prior to this technological development. Combustion leaves residue in the barrel of your firearm, no combustion… no residue that at times can be a hassle to clean.

Lastly, this innovation would allow "any" type of explosive payload to be used as a projectile, silently, simply because there is no physical interaction between the firing mechanism and the projectile. This benefit would allow scientist and the military to explore more projectile options that would save more lives of our armed forces and bring more "heat" to the dangerous threats of terrorist groups across the world. We would see this "IDea" enhance our soldiers tremendously. From any and all hand held weaponry to with "unbeatable" rates of fire power, to aircraft and ground vehicles. Every type of of projectile weapon to be enhanced towards its final innovation.

IDea, 63 Magnetic Gears

Going back towards IDea, 57 and IDea, 58 and the use of magnetism for transportation. I develop in more detail on a different design, which brings me to IDea, 63 Magnetic Gears. Which does include some mathematics that may or may not be accurate or relevant towards the goal of my design.

Idea, 63, Magnetic Gears.

Idea, 7

1 Gear

inspiration Idea, 57.

bikes, cars, 4 wheelers, snow machines, Tanks. anything with tires or tracks.

500 lbs per tube when all tubes are going.

gears being activated & deactivated from inside out.

1,000 lbs of instant thrust.

1,000 lbs of constant thrust.

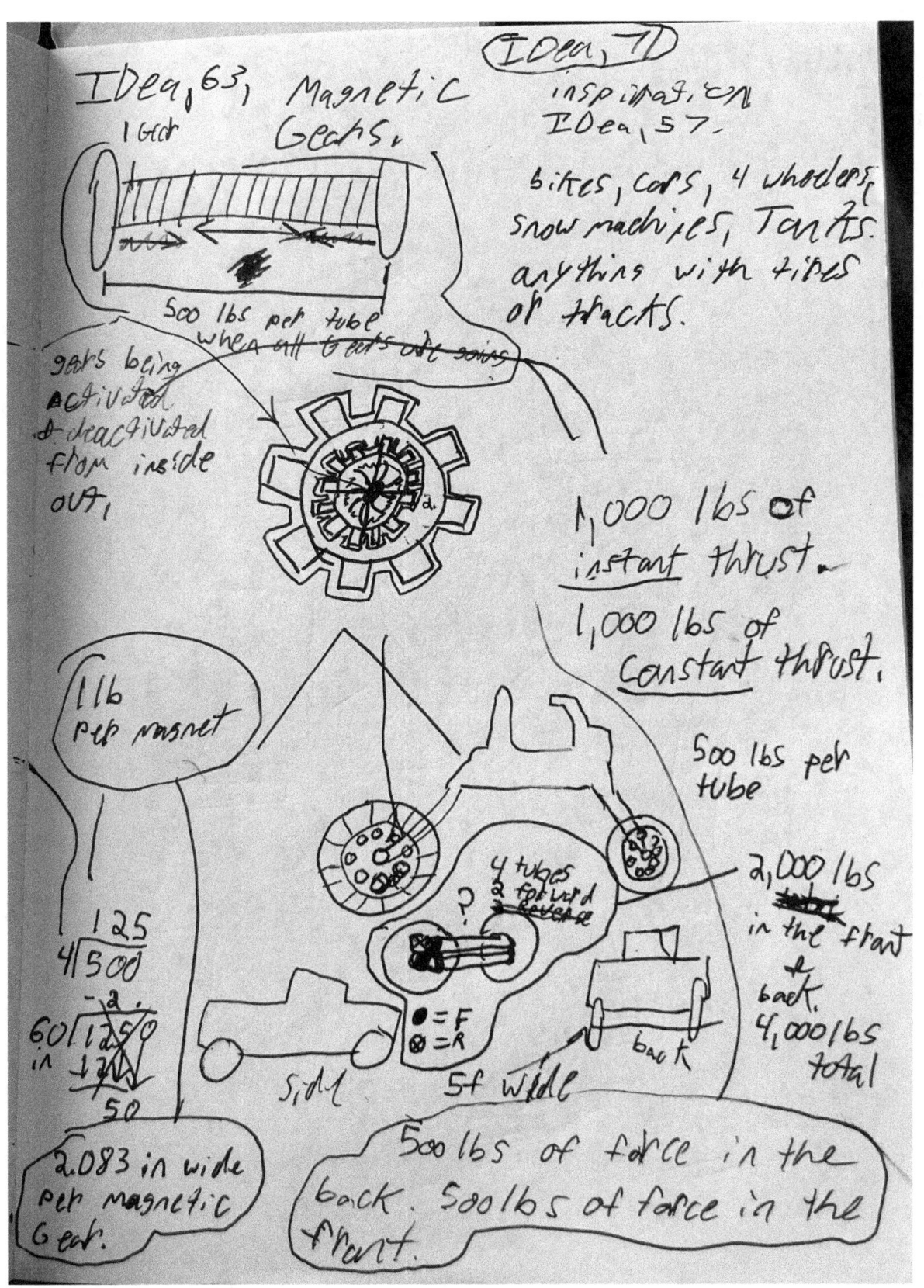

1 lb per magnet

500 lbs per tube

2,000 lbs in the front & back. 4,000 lbs total

4 tubes 2 forward 2 reverse
● = F
⊗ = R

side
5ft wide
back

125
4|500

60|1250

50

2.083 in wide per magnetic Gear.

500 lbs of force in the back. 500 lbs of force in the front.

```
        .4166
      6)2.5000
       -24
        10
        -6
         40
        -36
          40
         -36
            
                          1
   .1041⁵          2.46640
       2              x  2
   .20830         4.93280
       2
   .41660
       2
   .83320
       2
  2.46640

  5lbs
    x
    2
```

```
   .1041⁵
  4).41660
    -4
    016
    -16
      06
      -4
       20
      -20
       0
```

outer gear (#2) would electrically give 4 pounds of resistance to the 4 pounds of magnetic force, the brakes

2 wheels

2.5 lbs | 2.5 lbs

6 magnetic gears

.4166 lbs per gear

4 magnets per gear

.1041⁵ pounds per magnet

10 lbs → .20830 pounds per magnet
20 lbs → .41660 pounds per magnet
40 lbs → 1.23320 pounds per magnet
80 lbs → 2.46640 pounds per magnet
160 lbs → 4.93280 pounds per magnet

the width of a tire could support layers of magnetic gears.

I still think about this "IDea" to this day. Can you imagine immense power, free electrical power, and a lighter vehicle all in one? This "IDea" would accomplish all of that. First of all, lets take the heavy engine out. Then remove about 70% of the mechanical "guts" of the vehicle, there is no need for them anymore. There is no need for transmission oil or cooling fluid, we can leave the windshield wiper fluid though, that's fine. Replace the drive train with metal tubes containing the magnetic gears. Also, replace the tires and rims with the magnetically geared tires. Last, but not least, add some electronics to "sync" and "command" all of the gears that will register with drivers physical commands.

Then, with a revolutionizing step forward, the new car is powered on and 1,500 pounds of instant and constant thrust is at the drivers command. Add some magnetic alternators and magnetic generators, then the driver driver can venture much greater distances than "any" innovation before.

Imagine a perpetual motion machine being magnetic and generating large amounts of power, the impossible became possible. This innovation is very exciting with the potential it offers. This "will" completely upgrade, improve and revolutionize the modern day "wheel and track". If you create this, you would own the new industry of vehicular creations.

Now, how would I make all of these magnetic innovations better? From this point forward you, the reader, will witness a transformation. An evolution from basic theory to theories of revolutionary proportions. The way my "IDea's" escalated and took shape is an honest result of an accident. I remember being woken up in the middle of the night with an "IDea".

It would appear in on me with nothing that would have sparked these thoughts in my subconscious. However, the moment it would arrive would be the moment it would manifest my thoughts, almost driving me "mad". Being overwhelmed with excitement, I would find myself in disbelief that it was I who was given this responsibility. How would I approach these "mysteries of the night" that would "haunt" me with excitement and intellectual grief? I have no college degrees and yet I bared these thoughts of innovation. It was in my power to write down, the best I could, the encounters I went through.

IDea, 68 & IDea, 65 To Measure Magnetic Force

With everything being said and confessed I would like to take us back to the question that would take me on a scientific journey and open doors to "IDea's" and theories that never crossed my mind or my world observation. How would I make these magnetic innovations better? That question brings me to "IDea, 68", the very first looks into that question that would later on inspire many "IDea's". Those "IDea's" will be later found in this book.

> IDea, 68.
>
> Goal: electrically enhance magnetic force.
>
> Theory: magnet is a mixture of metallic sillica. would adding <u>conductive</u> sillica to the mixture allow the magnet to except electricity? would it enhance the magnets force?

Then I hit a mental wall, temporarily. I began doing small home science experiments. I used a kitchen scale and two DIY Magneodymium Disk Magnets. I wanted to see how much strength the magnets possessed. I am going to take us back a little bit, to "IDea, 65 To Measure Magnetic Force". When starting the experiments I had high hopes. However, I would soon be puzzled and at a loss for further efforts.

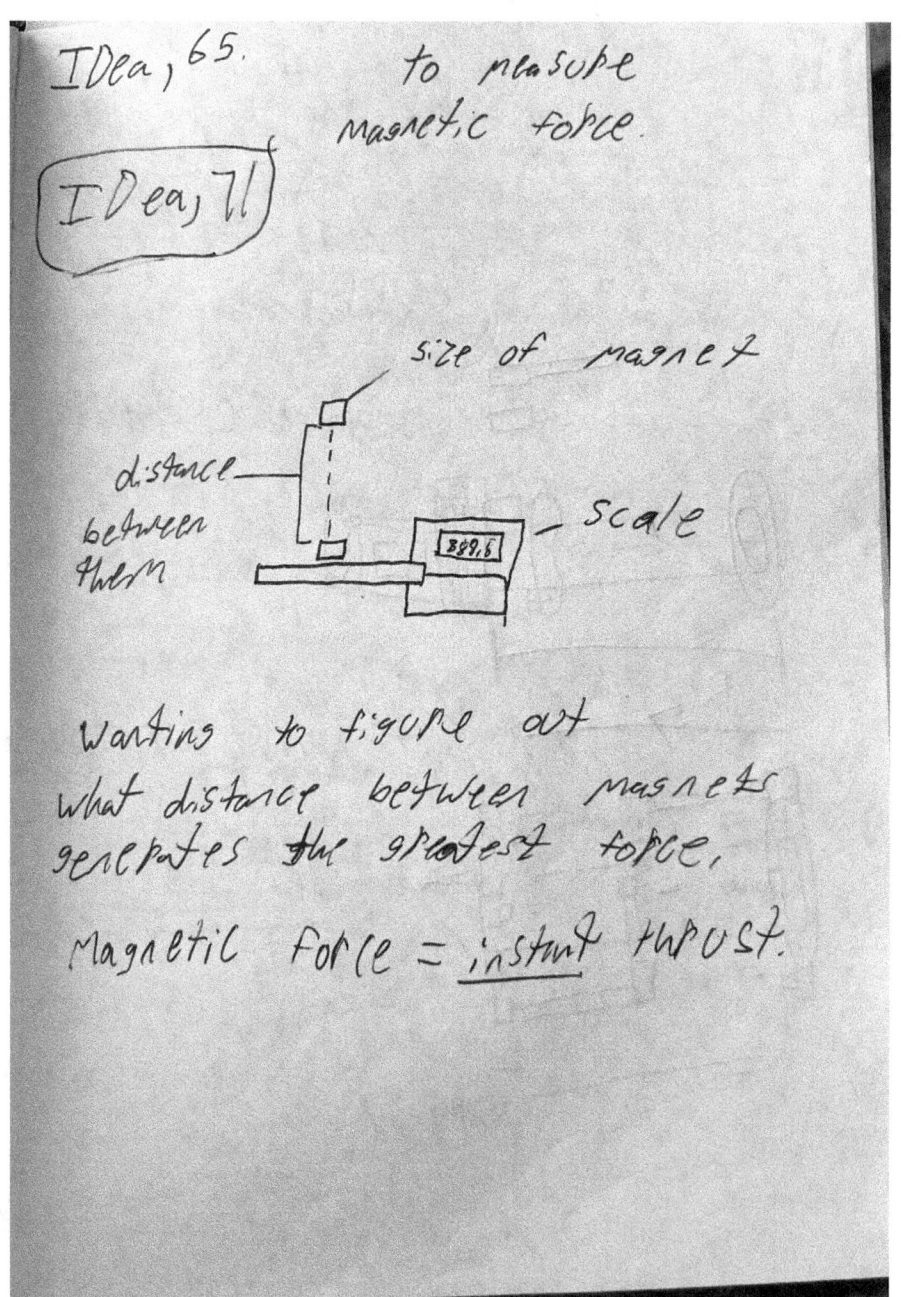

DIY Mag Neodymium disk Magnets.

1.26" x 1/8"
distance
1 lb = 1 mm
0 lb = 3 in
weightless

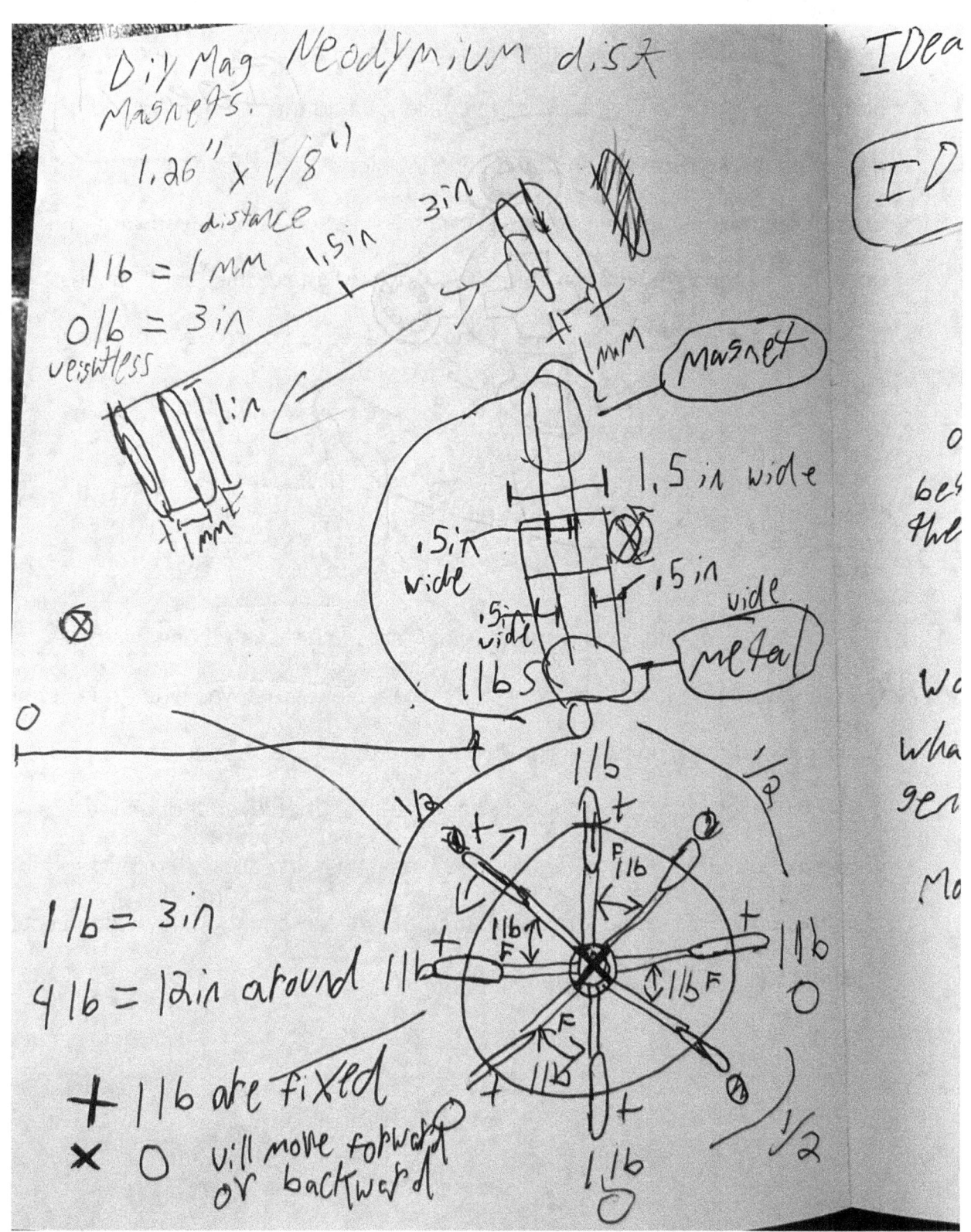

1 lb = 3 in
4 lb = 12 in around 1 lb

+ 1 lb are fixed
× 0 will move forward or backward

I thought I was onto something. However, with the magnetic experiment I found that the scale gave me the strength of the magnet. But, trying to "craft" a demonstration would be the puzzling challenge and the "mental wall" I hit. I took a cube shaped Styrofoam and made two incisions leaving one millimeter of Styrofoam between them. I put a disk magnet in each slit. I played with the positive and negative sides to figure out, What will make this "light" piece of Styrofoam move? But, nothing happened… there was no movement what so ever.

IDea, 71

I determined that one millimeter of distance between these two disk magnets would generate one pound of force, at least that was what the kitchen scale showed me. The two disk magnets and the piece of Styrofoam equally weigh much less that one pound. I shortly realized that, even though the scale showed one pound of force, equal force was also being applied to my hand that was holding the disk magnet against the other disk magnet on the kitchen scale. This realization would bring me to "IDea, 71"

Idea 71.

All previous IDea's regarding magnetism, magnet force on magnet wont work, magnetic force on metal might work.

magnet

piece of metal

?

magnetic geck

teeth along tunnel

One full rotation of interior metal for 1 turn of gear

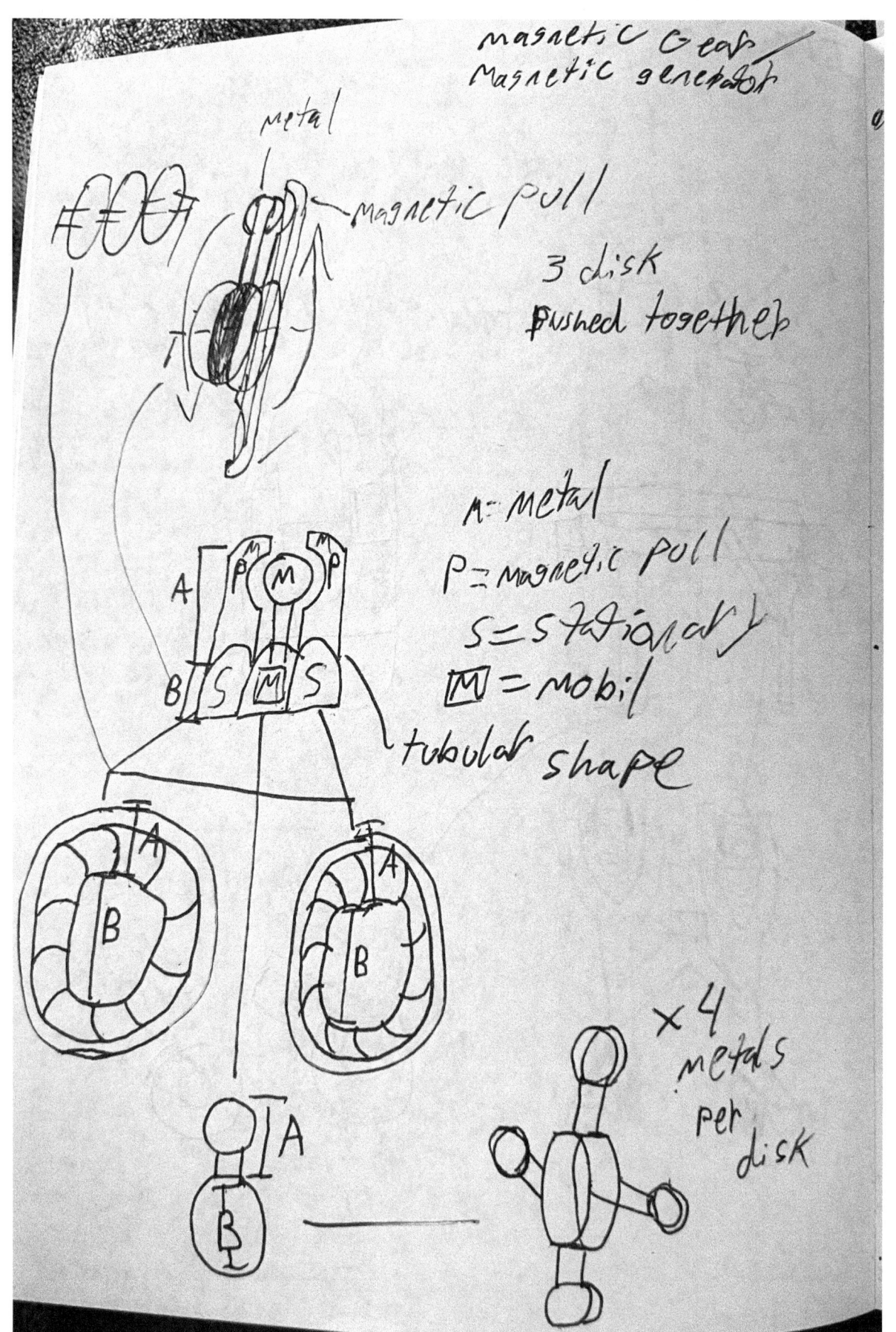

IDea, 78

The "mental wall" I hit may not have been such a problem after all. I did not proceed with any physical experiments. Which looking back, I could have tried to use the disk magnet and a piece of metal to move the Styrofoam independently. I stuck to brainstorming every possible angle or application. Adjustments and alterations would need to take place with my magnetic "IDea's" prior. With this much needed "basic" understanding, I smiled and allowed myself to be excited once again. There was still hope for my "IDea's". Amongst the adjustments and alterations was "IDea, 78 Magnetic Pattern".

Idea, 78. Refer back to Idea, 71, & Idea, 59, 57, 58, 63, 65

For magnetic pull to be accomplished, A magnetic pattern must be in place.

metal / magnets (diagram)

if both "Mgs" are equally attracting the "metal" It will "halt" or stop the metal.

(zigzag pattern diagram)

by the "pattern" placement of magnets. It pulls the metal forward from Left to Right.

This is where I stop with any magnetic pursuit, for now. I am now going to dive into something completely different than what was said in the book so far. However, a surprising connection will link between the pages you have read to the pages unopened. Please bare with me and keep your mind open.

IDea, 79 Teleportation

The next topic is my first thought of the already well known idea of teleportation. Please sit back down, open this page, and hear me out. I am going to throw science "fiction" at you, but I challenge your minds to embrace the possible science "non-fiction". These are my sincere thoughts of exploration towards the examples of theoretical physics. I wanted to take a shot at supplying the "possible" answers for these scientific mysteries. Later in this book, you will find that not only the evolution of my "IDea's" are compelling but they are a lot closer to reality than we think. Lets begin with "IDea, 79 Teleportation"

Idea, 79. teleportation

```
Point              Point
 A  |=========|  B
        ↑ c.
```
↑ strings in fabric of space.

C. to manipulate this string to the shape of a type of wave or frequency

```
A |∿∿∿∿∿∿∿| B
      c.
```

by making the waves higher & more compressed would bring point B to point A.

```
A |))))| B
```
once at point B the surrounding strings are at the original distance in between points A & B. According to that distance would reflect whatever jump forward into time.

string C. would be the current time on earth.

other strings would be whatever light years away point B is than point A.

Applications for time travel into future.

In simpler terms, imagine this scene. There is a home with a door in the middle and two windows on both sides of that door. As you walk into that house, you go to either the left side or the right side. Then you look out of one of those windows to see the walking path you took from outside to this house.

Now, that I have painted this picture in your mind, let us include in the picture a a pot and a seed off of a tree. First, you put the seed into a pot, then you walk into the house, and the door remains open. You go to either the left side or the right side to look out one of the windows. You see that there is a fully grown tree where the pot and the seed use to be. Now, you walk back to the open door to look outside. You see the seed and the pot you placed it in, you understand?

I believe this is how teleportation and or time travel will work. Finally, let us turn the windows, that are on both the left and the right side of the door, into doors themselves. There are now three doors, total, for the house. You walk from the outside, following the same walking path from before, into the house. The difference now is instead of "looking" out of one of the two windows, you "walk" out of one of the two other doors. When you walk out of one of those doors you find a fully grown tree.

IDea, 102

However, "time" and the passing of time did not impact you to any degree what so ever. You are the same age going outside as you were entering the house. I follow up this theory with a simple "planetary" illustration that replaces the tree and the house, full of doors and windows, with the planet Earth and another planet that is 100 light years away. Let us begin with the illustration.

Idea, 102 — reference to Idea, 79, teleportation

1. O — earth 2000 years in future
2. O — Earth present day sound/manipulating waves → O 100 light years away.
3. O — Earth light 100 years future, look back → O
   n has already existed for 100 light years

- With sophisticated space travel, time travel is real. However, travel to the future is the only time period. Cannot travel to the past.

With sophisticated "space travel", "time travel" is possible and real. However, traveling to the future is the only direction of time we can travel. You can only travel back to the day when the technology was established and or used. If my theory theory is correct, "all" ancient mysteries can not be traveled, to be solved once and for all. Unless of course, we discover a different course of technology with "pre existing" time periods to travel to.

IDea, 110

Now, I will take you to "IDea, 110". This is where is question Albert Einstein's special theory of relativity. I love Albert Einstein and I believe he would have been proudly open to, listen to thoughts and theories, as an opportunity to teach and to enlighten those who may have disagreements of his methods. Additionally, I will take my theory of space travel and time travel to unravel certain events with Orions Belt. This is where my teleportation "IDea" meets the "Ancient Astronaut Theory". Let us begin with the questioning of Albert Einstein.

Idea, 110.

$(Time = distance = Space)$

Reference Idea, 79 + Idea, 102.

Einsteins special theory of relativity
space time

? wouldn't that just create a "black hole"? any amount of "mass" times a number as high as speed of light... would be "insane"

$E = MC^2$

Energy = mass × speed of light²
(in this perspective his equation is flawed)

orions belt

. . . — Mintaka 1200 light years

Alnitak 8/7 light years

Alnilam 1340 light years

Now, I am sure I might be wrong. I will admit that I am not too academically "sound" when it comes to questioning a great genius of a scientist that is Albert Einstein. However, I do reserve the right to question what does not quite make since to me. Maybe the perspective I bring up is flawed, but it is interesting to ponder. Now, I will take us to the ancient connection part of IDea, 110.

1353 bc
egyptian pharaoh Amenhotep IV
adopts a new deity, Aten, &
changes his name to Akhenaten.
worship of the sun.

4 bc
Birth of Jesus Christ (son of God)

1412
Inca empire worship of
Inti (diety of the sun)
Pachacuti Inca Yupanqui.
"Son of the sun"

(Alnilam
1340 light years
away)

(Alnitak 817 light years away)

9th century BC   900 BC – 801 BC
mesoamerican people built pyramids
from 1000 B.C & 400 B.C 1st
pyramid built by the Olmecs

teotihuacán built the pyramids of
the sun & the moon between
1 A.D. & 250 A.D.

(Mintaka 1200 light years away)

1200 BC
1st civilization in Central & North
America. Known as the Olmec
civilization.

According to the "Ancient Astronaut Theory", our Ancient Ancestors were influenced by extra terrestrials. What if it was true? If we are assuming that our Ancestors were visited by "beings" from Outer Space, that means they possessed sophisticated space travel. They were able to leave there footprints and the same "agenda" throughout different time periods. How is it that cultures and civilizations, separated by vast distances and vast ages, are able to construct buildings and structures very similar to each other.

IDea, 138 Cross Cultural Examination Equation

The answer is this, the same architect, throughout different time periods in history, different locations throughout parts of the world, and three different worships of the "same" God or Deity. When I made this connection, things clicked and made complete since to me. However, this is just a theory, a theory built on deductive reasoning or quite simply coincidental. I did put a name on that process I used to create my conclusion. That takes the form of "IDea, 138 Cross Cultural Examination Equation"

IDea, 138. Cross Cultural examination equation.

CCEE A Reference term (Adj)
to use and or refer to different cultures, from any time period (past or present), for their technological achievments, the uniqueness of their cultural lifestyle, and their challenges that they overcame, to solve a problem, to solve a mystery, to overcome present day challenges, and to aid in technological discovery.

Introduction to Magnetic Manipulation

Now, I want to bring us back to Magnetism. I promise, my readers, that "everything" I have mentioned will start piecing together. I hope you have been enjoying this pursuit. To ponder the possibilities that contain a shadow of doubt to it factuality and certainty. Ladies and gentlemen, you have been very patient with me, thank you for that. For this last chapter, I promise that you will walk away from this book with new thoughts and new ideas. This next "IDea" is one that takes magnetic forces to the next level, that the world has never seen yet. I present my first, of many, thoughts into Magnetic Manipulation.

IDea, 114

Idea, 114 scattered circuitry board within magnet. ability to electronically activate/disactivate a magnet and electrically enhance magnetic force.

---

in addition

the utilization of electronics, would be able to change the shape of magnetic field. focus magnetic force & energy to slightest particles/atoms. small electronic magnet + power source + electronics magnetic propulsion/magnetic manipulation of objects.

(N/E/S) Earth has 2 Pols, has Magnetic field.

(N/S) magnets has 2 Pols has magnetic field

Earth has gravity Magnets has gravity? Application possibly artificial gravity

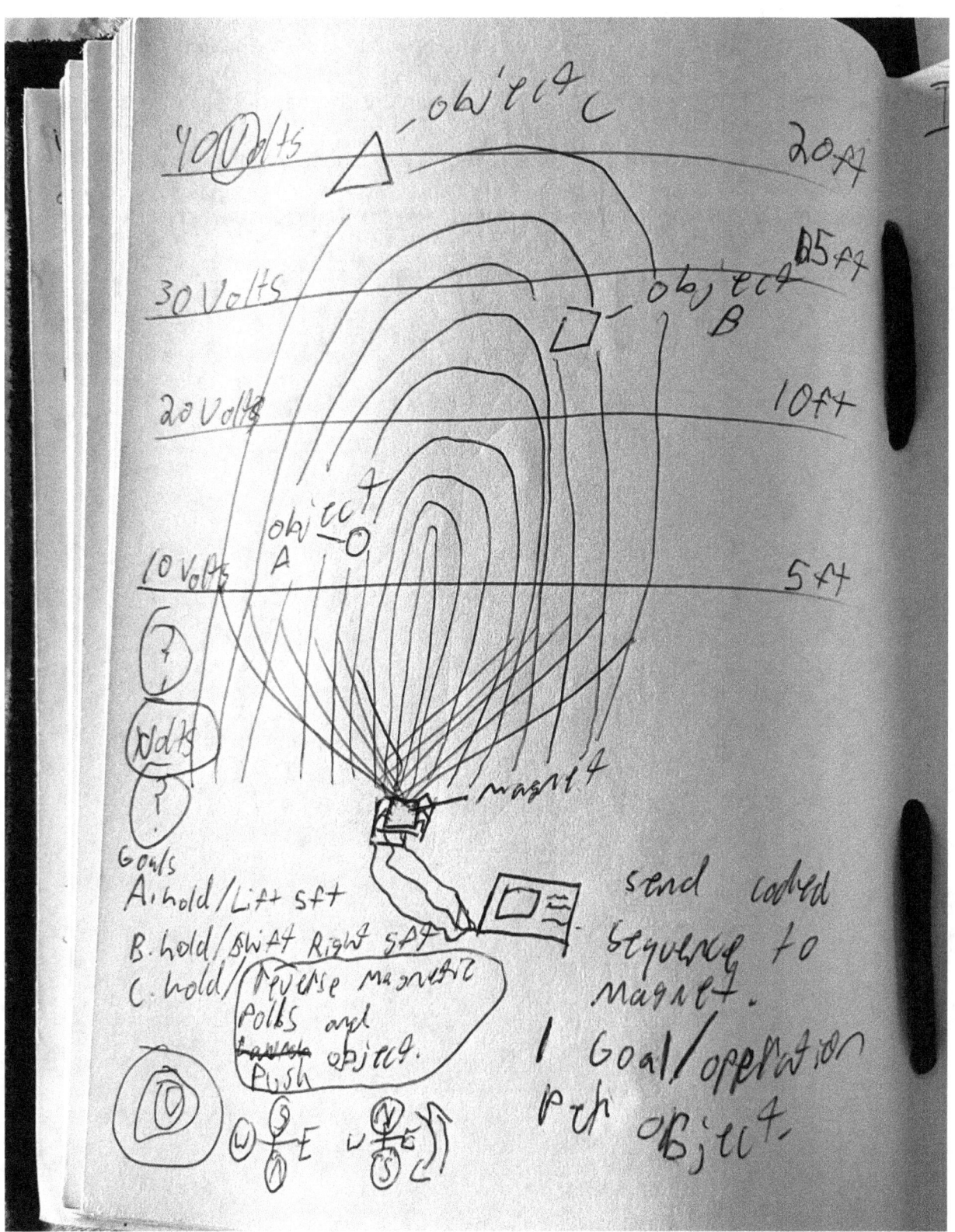

Goals
A. hold/Lift 5ft
B. hold/Shift Right 5ft
C. hold/reverse magnetic poles and Push object.

send coded sequence to magnet.

1 Goal/operation per object.

Let us take the ordinary magnet to the next level. To harness the ability to command and program a magnet would revolutionize the way new view magnetism. This innovation "will" grow mankind tremendously. I personally would love to see what "language" magnets possess when plugged into a computer. I think we will discover the answer to that question in our near future. I will now expand Magnetic Manipulation with IDea, 133.

IDea, 133

Idea, 133 in addition to Idea, 114.

As soon as we successfully connect our special magnet with a computer we can then begin running some tests. When another magnet goes in front of our SPM at what point will our computer detect the magnetic presence of that magnet.

By collecting enough data we will be able to analyze/collect info of objects magnetic be able to solve how much mass a object has, what amount of energy to enhance the SPM to electric ~~field~~ manipulate object.

⊕ positive ⊖ negative residue exist in all matter.

Be able to magnify the magnetic field to spm disturb/control/manipulate the negative and or positive particles that exist in all matter.

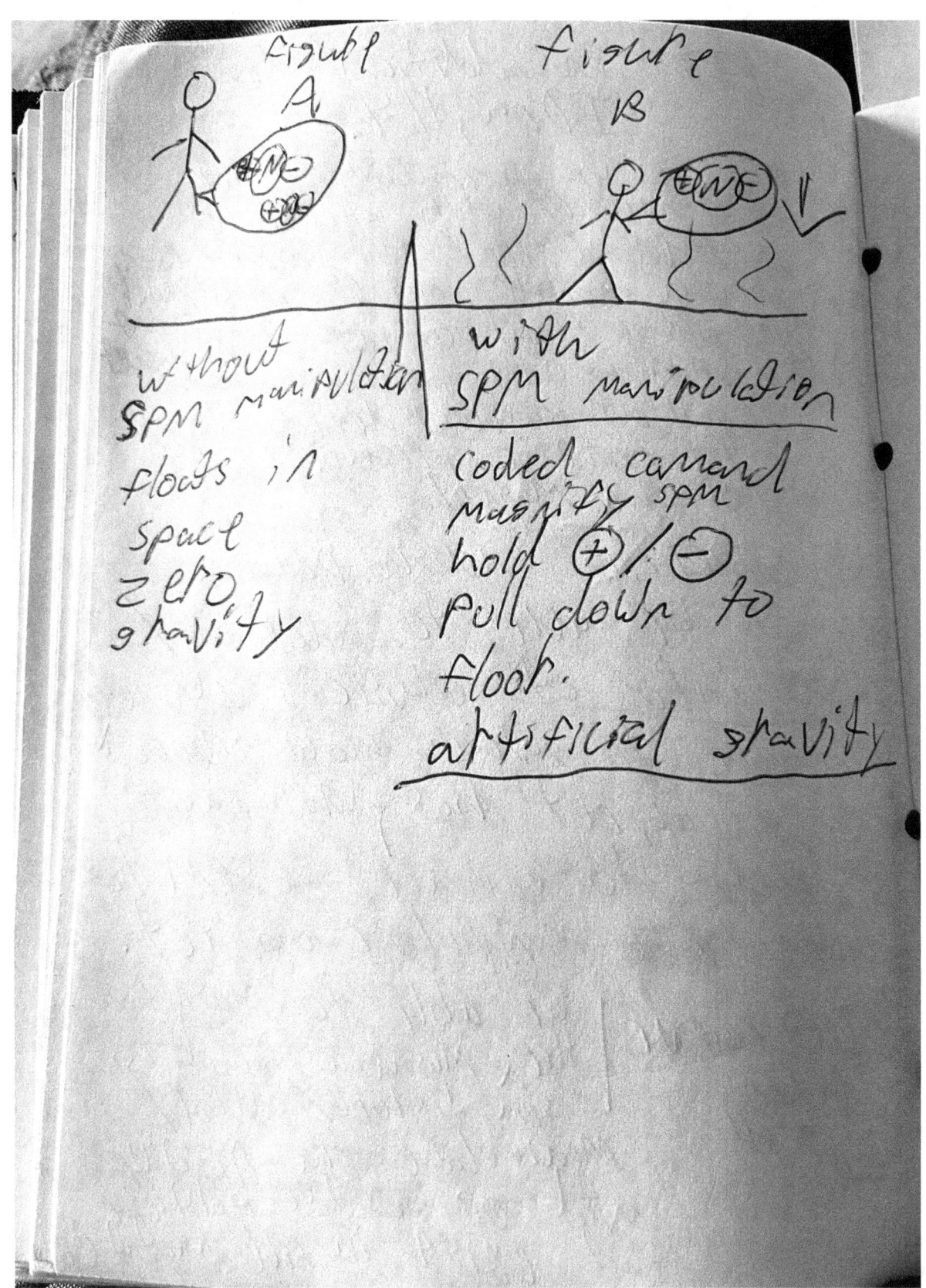

figure A | figure B

without SPM manipulation | with SPM manipulation

floats in space zero gravity | coded command magnify SPM hold ⊕/⊖ pull down to floor. artificial gravity

artificial black holes?
wormholes?
ability to bend the fabric of space + time using
(SPM)

Earths atmosphere is layers or ripples of Earths magnetic field.
gravity < reflects - atmosphere.
         equals - atmosphere

tighter or stronger the gravity equals layers/strength of atmosphere.
coinsidence?
the Moon little to no gravity and zero atmosphere.
Mars little to no gravity and little to no atmosphere.

With my theory, I began two have a greater appreciation of our planet Earth. With everything already known of Earth's unique qualities to create and preserve life, I uncovered more of Earth's unique and "precise" qualities. Earth's atmosphere is more complex than one may think. According to my theory, Earth's magnetic field is also responsible for our atmosphere for it keeps our most precious gases close to the Earth's surface, like oxygen for example. The further away we get from Earth's surface, the less oxygen you have.

Earth's natural "program" ,within its magnetic field, allows for this unique and precise manipulation of oxygen. Earth's magnetic field could have picked and aligned differently any other type of present gas and our world would be completely different. This realization made me appreciate, tremendously, the planet we call home.

However, it also made me realize that we must take into consideration the magnetic field's on Earth like planets. For they might be "naturally" programed to be much different than ours. This is a very important detail we must think about when identifying Earth like planets. With planetary atmospheres aside, I want to get into defying the Laws of Physics, via Magnetic Manipulation, with "IDea, 140".

IDea, 140

Idea, 140 Magnetic manipulation to repel every universal influence (to defy laws of physics)

no influence = worm hole a "teleportation" tear in space + time

(true invisibility)

[Answer to Fight global warming]

Refer to Idea, 110

A stars constant energy equals the light constantly being emitted

light is emitted like a river not like rain drops.

"All stars are linked + connected by light."

☆ ——light———•Earth

light is a constant only if A "Living" stars energy is a constant.

Idea, 114   Idea, 110
Idea, 133   Idea, 140
Idea, 79   Idea, 102

Constant energy = speed of light × distance light cov[ers]
   ↑ stars   constant   constant
Constant energy = constant speed of light • constant dist[ance]
"(E)" power of 2 equals "constant" | Sansen's special
                                     theory of relativity
$E^2 = C^2 \cdot D^2$ $\boxed{E^2 = D^2 C^2}$

Q if only solar eclipses happened more often

Q blocking out suns rays via magnetic manipulation

Space travel is "lonely" without "teleportation" traveling speed of light will get you to the stars "Alive" but, with no teleportation there is no returning to the time period when you left.... But this route would be ideal for "time travel".

If light is a _constant_... How can we say that depending on the distance of a star equals how long it took for the light we see, to reach us? ~~are constants timeless in regards to the age of a star?~~

Is a constant light timeless in regards to the age of a star?

if a star suddenly dies (star loses its energy) the ~~~~ constant light would no longer be emitted... like a flashlight when it runs out of battery.

the light will _instantly_ turn off. what are constants about "stars"

Distance from earth, energy, light speed of light equals 300,000 km/sec

IDea, 143 Time Travel Through Space Travel

With this "IDea" being said, does this mean that stars would represent "coordinates" for means of time travel? Astronomy has already made it easy by organizing some of the stars into horoscopes. Each horoscope would contain different locations and different times reflected back on the time period on Earth. With so much structure already in place, it gives one hope and confidence to one day venture deep into space travel and time travel.

IOla, 143 time travel through space travel.

For every "light year" of distance you travel equals a year into the future.

A. 2020 Earth — teleportation or traveling speed) equal of light → B. 200 light years away / star

traveling forward

A. Earth ← ——————— B.
2020

This equation works by leaving point A. traveling to point B. only to return to point A.

IDea, 100 Parallel Universes & Our Mind

The Connection

With all of this talk about space travel and time travel, do I dare consider the possibility of traveling through parallel universes? Before the prospects of traveling through parallel universes, we would have to accept the possibility of parallel universes themselves. The theory already exists and it has existed for sometime. Now, I have an "IDea" that, not only supports the existence of parallel universes, but my "IDea" offers different perspectives and clues towards further investigations to proving the "existence" of parallel universes. With enough talk, let me present to you "IDea, 100 Parallel Universes & Our Mind…The Connection".

Idea, 100. Parrallel Universes
& our Mind.
the connection.

1. <u>our Mind</u>.
   1. memories we have
   2. fears we have
   3. characters we know
   4. environments we know
   5. tramatic experiences we know
   6. info our 5 senses has gathered. ex (eyes)
   7. 10% brain used.

2. <u>Mysterious connections we live</u>
   1. Deja vou
   2. instincts
   3. gut feelings
   4. "little voice in your head."
   5. "bad feelings about this"

3. <u>Mysterious connections we "Dream"</u>
   1. stats/info about how many times we dream a night.
   2. strange realistic dreams
      1. dreams we don't <u>recognize</u> →
      2. No familiar info   continued
      3. first person dreaming

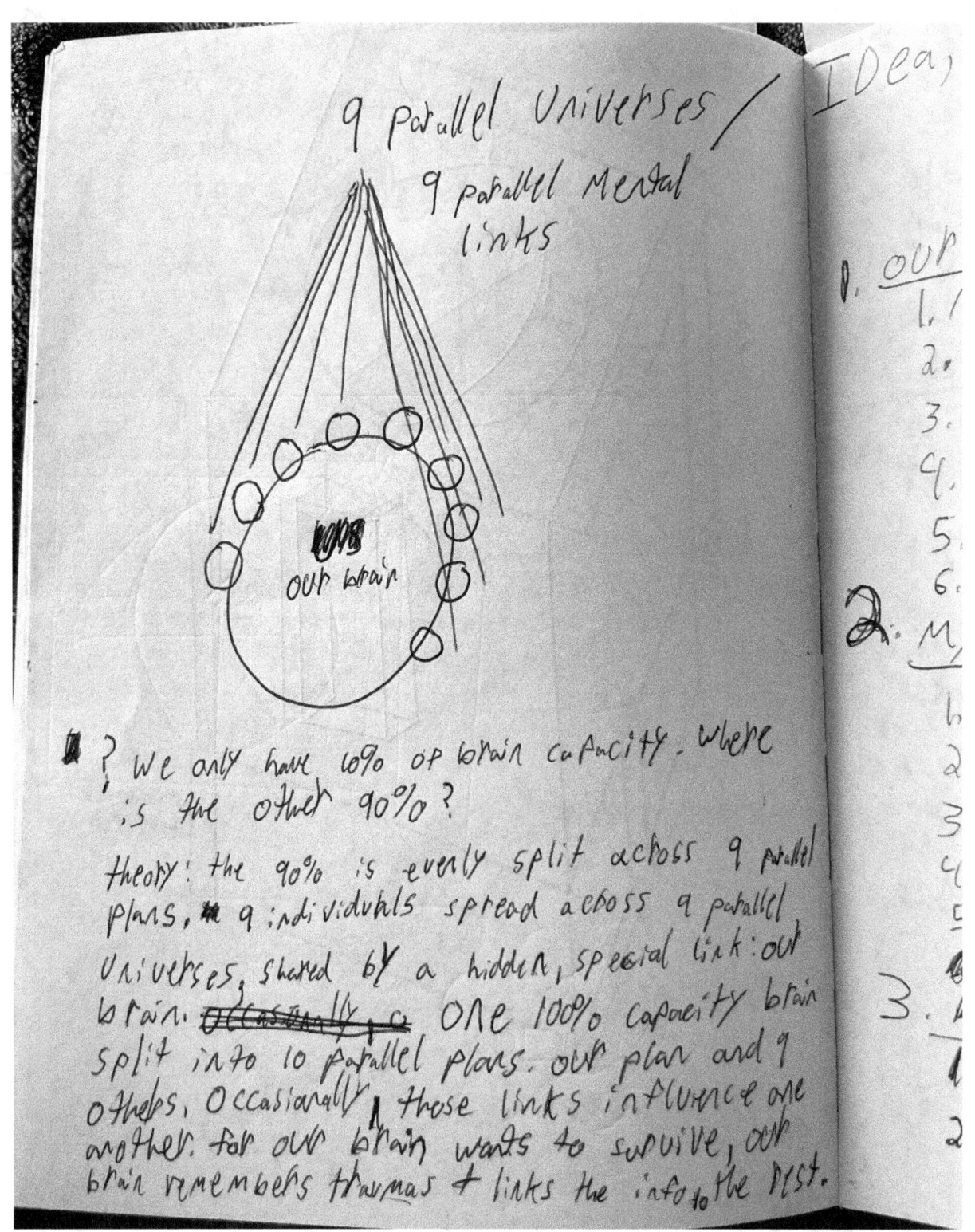

? We only have 10% of brain capacity. Where is the other 90%?

theory: the 90% is evenly split across 9 parallel plans, 9 individuals spread across 9 parallel universes, shared by a hidden, special link: our brain. ~~occasionally~~ one 100% capacity brain split into 10 parallel plans: our plan and 9 others. Occasionally, those links influence one another. for our brain wants to survive, our brain remembers traumas & links the info to the rest.

> 3. question? A person born blind, do they have dreams with colored environments, characters of people? if yes to any of these questions, evidence of this connection?

First lets take a look at the diagram of our brain. Our brain contains an immense amount of information that is gathered from our 5 senses. This information leads to the memories we have, the fears we have, the people and characters we know, the environments we know, and the traumatic experiences we know and lived through.

In life we encounter many strange phenomenas like deja vu, instincts, gut feelings, the little voice inside your head, and moments where you say "I have a bad feeling about this". Is it possible that, in a different parallel universe, we have experienced those bad times? I believe it is very possible if not undoubtedly true. Have you ever seen something before it happens? And before those events take place, you choose to quickly walk a different path or walk at a faster pace?

I have my experiences like that and I have witnessed a change of "fate". That is what prompted me to think deeply on, what happened and why? The result became "IDea, 100" which made me question self proclaimed psychics and or "witch doctors". Where the few have more "insights" than others. I expand my thoughts more on the 10 parallel universes and put them in order as if it is some kind of time line.

0 0 0 0 0 0 0 0 0

1.   5.   10.

Additionally, I continue my theory by saying that each individual person are in there own distinctive position on that chart. Those that are accident prone or "unlucky" you might say, would be between the 1 and the 5, due to the decreased level of "insight" and or the decreased level of "gut feeling". Those that do not have that many accidents or those who are "lucky" you might say, would be between the 5 and the 10, due to the higher levels of "insight" and or higher levels of "gut feeling". You can logically determine that psychics or the famous foretellers are closer to the 10 on those charts due to there certain "acute insight" abilities and or visions. Of course, in the ancient and historical times, those who might have seen devastation coming there way would have little to know protection thus succumbing to what is deemed as "fate".

The last connection is also one that we live with, that is our dreams. The rare and incredibly bizarre dreams we have that that make absolutely no logical since, with the information our mind has gathered over the years of our existence. I am talking about the dreams that are in first person perspective, in an area that is not familiar, not even in the slightest. Dreams that we do not have any control over, gives the feeling of "going on a ride" or "hands off of the wheel" dream experience.

Additionally, these dreams are not scary, haunting, or even chilling to bare. There does not appear to be much "emotion" from our perspective, confusion most likely, just a ride and we are the observers. My theory is that those "specific" dreams are indeed connections from our universe to a parallel one. The explanation for the bizarre nature is because the world we are seeing and experiencing in those dreams are nothing that out "5 senses" have gathered here in our universe. The next time you have a very odd and or strange dream, you just might be looking through the eyes of a parallel "you".

Before I close this "IDea" I must ask you another question? For a person that is born blind, do they have dreams with "colored" environments with characters and people? If the answer is yes, is this "evidence" of the connection from our world and the parallel ones?

IDea, 144

I "think" we have covered all of the "travel" possibilities using magnetic manipulation. I am now going to present you my next expansion of Magnetic Manipulation with "IDea, 144". This is also an introduction towards the concept of Magnetic Filtration, which would technologically satisfy the modern era.

Idea, 144 — Magnetic manipulation to condense or create clusters of particles and energy.

Never been seen before on earth... never been created by man befor.

What would the result be if we condensed

- light?
- electricity?
- air or oxygen?
- types of gases?
- fire?

Weaponized applications.

to harness the power of nature via magnetic manipulation.

New ways of travel application

Ultimate electrical power efficiency

thus not allowing air bubbles to escape. imagin an air bobble in a bottle that is under water. the magnetic field would filter in oxygen from the water, then filter out the water.

bottom of the ocean

A magnetic field to filter in light. or to filter in heat sygnoters

the ~~man~~ inside this person

field would be able to look out through this field like a camera lens.

"IDea, 144 Magnetic Manipulation to condense and or create clusters of particles and energy. Never been seen before on Earth. Never been created by anyone before. What would the result be if we condensed "light"? Or electricity? Or fire? Or oxygen? Or any type of gas?

Militarized application.

To harness the power of nature via Magnetic Manipulation.

New ways of travel application. Particle bridges that would electronically turn on and activate.

Ultimate electrical power efficiency.

Then to use Magnetic Manipulation to "magnetically filter" in or out any gas, virus, bacteria, particle, or matter."

And lastly, an example of fire extinguishing via Magnetic Filtration to filter "out" oxygen from a flame. Imagine satellites in the sky with this application. Or a less extreme, towers on the ground with this application. With the authorization of state or local firefighters, this would put an end to lives being taken away. Not to mention the thousands of homes and businesses lost "annually". The application would save local and and state governments millions of dollars, "annually". This would be the "game changer" on several levels.

IDea, 149 True Hologram Technology Through Magnetic Manipulation & Magnetic Filtration

During the time I was writing the rough draft of this book, I had yet another "IDea". I felt it was perfect to include at this time in the book. Ladies and Gentleman I present to you "my" take on the hologram concept. The "IDea" made me realize how much is truly possible with the use of Magnetic Manipulation and Magnetic Filtration. Let me jump right into it with "IDea, 149".

Idea, 149 — true hologram technology through magnetic manipulation

(light) → [prism] → light spectrum

Δ0
words
messages
colors

(light)

Magnetic field altered by magnetic manipulation

We've covered that magnetic manipulation can generate true invisibility by blocking out "light" from an area. Now lets filter in specific patters of color and black out the rest of the spectrum of light. by rendering a magnetic field we can create hidden shapes, words, messages, and co colors, photos, videos, much like a CAD CAM software for 3D printing or laser cutting, that with the presence of "light" would then reveal the designs within the magnetic field. This technology would create the world most crisp and detailed television by reaching down to the molecular scales of light. This is the answer to the Hologram theory. I figured out the birth of this concept.

"We have covered that Magnetic Manipulation and Magnetic Filtration can generate true invisibility by blocking out (or bending) "light" from an area. Now, lets filter "in" specific patterns of color and block "out" the rest of the spectrum of light. By rendering a magnetic field, we can create hidden shapes, words, massages, colors, photos, videos, much like a CAD CAM software for 3D printing or laser cutting, that with the presence of "light" would then reveal the designs within the magnetic field. This technology would create the worlds most crisp and detailed television by reaching down to the molecular scale of light, but maybe T.V's would just be holograms at that point. This is the answer to the Hologram Theory. I figured out the steps required to bring to life this concept".

IDea, 151 Hologram Smart Hand-Held Devise

Now that I have "broken the ice" of hologram technology, I take it a step further. Let us apply this technology to mankind's most popular and personal devise. Let us bring hologram technology to a cell phone. I bring to you IDea, 151 Hologram Smart Hand-Held Devise.

Idea, 151

refer back
Idea, 149

[Sketch of handheld device with labels:]
- invisible magnetic grid
- ← Front
- hologram smart handheld devise
- actual physical devise
- charging port
- speaker
- main home button
- magnetic manipulation

By using magnetic manipulation, we can detect the users finger to select the option on the hologram's interface.

When we press the main button on the devise, the light is turned on along with the magnetic field thus creating the "hologram". The interface would be much like that of any modern day smart devise.

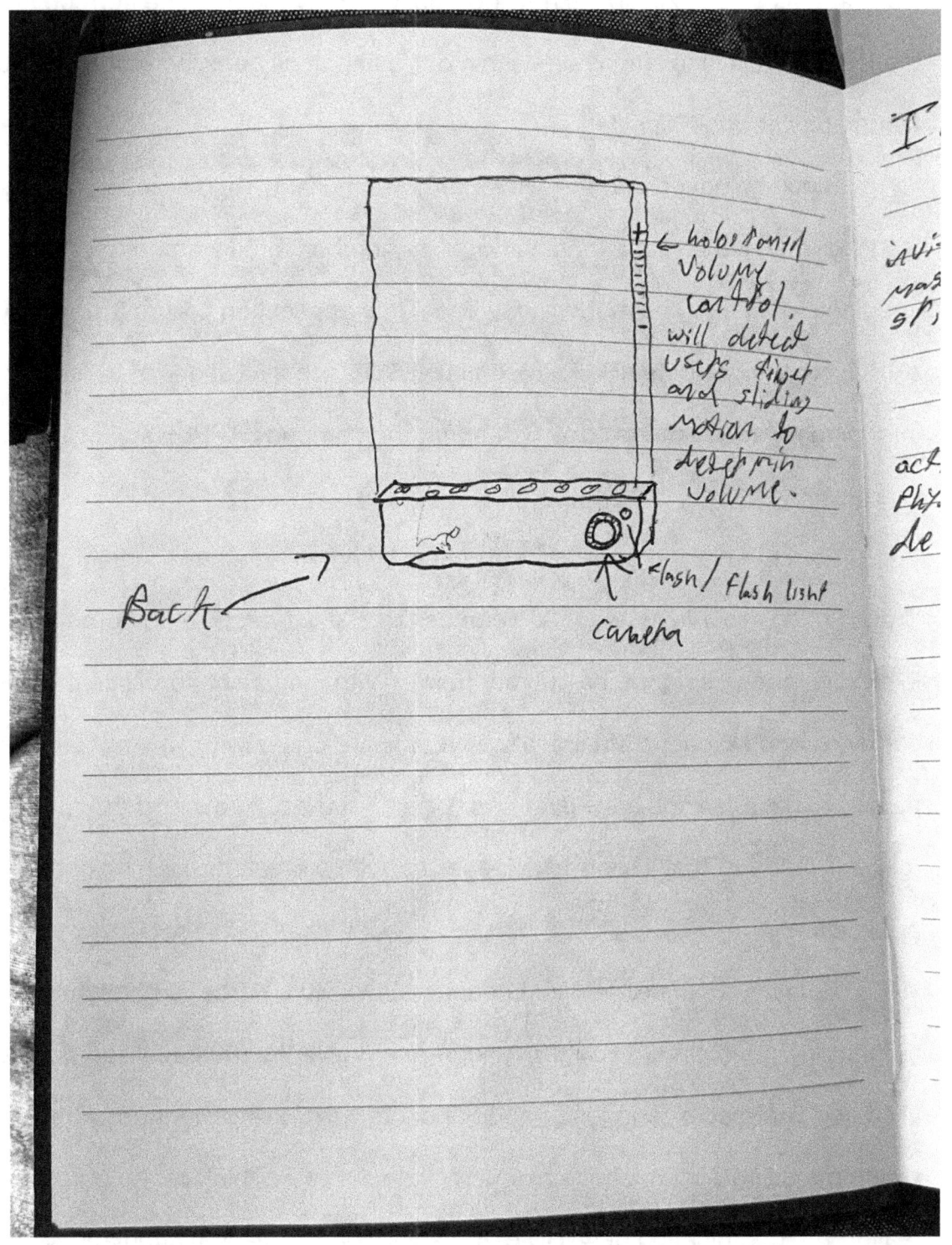

This technology, without a doubt, will revolutionize mankind. What once was seen as science fiction is now non-fiction. If my theories are correct, all of these wondrous innovations are within our grasp. If all goes incredibly well, this technology is within our lifetime.

Since I plugged into the "Hologram Smart Hand-Held Devise", this also lead me to an expansion. There is no "IDea number" but I wanted to bring imagination to you. The modern day FaceTime application brought to you by APPLE for the use of their APPLE IPHONES. I Zachary Sanger is, in no way, infringing upon any of the copy rights that the company APPLE so lawfully possess).

For creative and scientific purposes, I want my readers to imagine "FaceTiming" your loved one. As soon as the call connects, a hologram of your loved one appears right there in your home. With the use of sophisticated 4KHD cameras, and Magnetic Manipulation, your loved one is with you in the most in person experience you have ever had. Later in the book I expand the "hyper realism" factor of that in person experience, with a break through hologram application.

Before I continue, we must change the way we perceive hologram technology. My (Zachary Sanger) hologram will give the most realistic picture, vivid and sharp color, shades of color and intelligent use of light and shadow within the hologram. You have heard of "virtual" reality, think of "hologram" reality. The combination of 4KHD cameras, Magnetic Manipulation and Magnetic

Filtration will do something unimaginable. It will create artificial "texture" for the hologram. Here is how we do it.

First the 4KHD cameras will capture very crisp "details" of many surfaces on objects, animals, minerals, and humans. If you zoom in on any surface you will find patterns of depth, height, and width. Some objects are "sharp" and others are "broad". So how do we recreate those patterns to create artificial texture? Magnetic Manipulation will generate a certain magnetic resistance to the "human touch" determined by the depth, height, and width of a surface.

Another determination, is the kind of surface were trying to magnetically recreate. Sophisticated software will take place within the 4KHD camera. To intelligently detect what kinds of surfaces are in the cameras view. Surfaces like cloth for example will have a "minimal" magnetic resistance, to human touch, to accommodate the elements of the cloth along with the woven pattern that Magnetic Manipulation will recreate.

Some surfaces will have a "bounce back" kind of interaction with human touch. Other surfaces, like a table, would have a higher magnetic resistance, to human touch, to give more artificial life to the hologram. The 4KHD camera will show the splits in the wood of the table along its surface that Magnetic Manipulation will render all of those details through magnetic resistance, to human touch.

This brings me back you "FaceTiming" your loved one. They appear through hologram, hyper realistic, with an artificial touch and texture. They can live hundreds of miles away, yet they are with you. Maybe one of your parents will prepare a signature "meal" for you, yes this is possible. Do not freak out, but they can open the fridge, pick up a chefs knife, chop vegetables, and cook anything. This hologram technology can make this happen, with 4KHD cameras, and Magnetic Manipulation.

IDea, 158 Artificial Wetness or Other Conditions Through Hologram Technology

Along with artificial touch and texture we just covered, I expand to environmental conditions. This expansion is important in the goal of complete hologram reality. From creating environmental situations to enjoying a waterfall in your living room. The applications of this expansion are vast. Let me present my next "IDea".

IDea/59 Artificial "wetness" or other
conditions through Holostam technology.

Refer to IDea/49.
IDea/52.

Low level of "resistance" that quickly
"decreases" and or "dissolves" on your
touch. Synthetic water like simulation.

From rain drops to waterfalls to open
ocean.

The only thing that (so far) Holostam technology
"cannot" do is simulate temperature. We can
simulate wind that could bring about cooler
temperatures... much like a fan manipulating the
air to "blast" air thus cooling you down.

IDea, 159 Artificial Speech of the Hologram Through Magnetic Manipulation

This "IDea" is one of my favorite hologram applications. When I was first brainstorming the hologram applications, I stumbled on a problem. Would the holograms be connected to a speaker? I thought this would go against pure hyper realism of the hologram. I was thinking, what else can Magnetic Manipulation do? This question would lead to this "IDea".

IDea,159   IDea,15a† IDea,158   Refer to IDea,149

Artificial "speech" through magnetic manipulation. this would be used along side of true Hologram technology.

"Suppose to be a mouth"

From the air tube at the back of the throat, to our inner cheeks, our tongue, our teeth, and lastly the shape of our mouth. All of these factors working together to create speech. Magnetic Manipulation I believe, can and will recreate this "orchestra" of mankinds anatomical behavior. The shape of our mouth will be

Artificially created as "resistance", resistance to air molecules. No need to reconstruct the lungs. Using magnetic manipulation, the back of our Holostones mouth would have a small area of magnetic manipulation with one goal, to attract air molecules then to force the air molecules down our artificial "canal" of "resistance". Then the magic of our "orchestra" kicks in. Think of this... imagine the workings of a trumpet.

You compress air down a shaft which hits the walls of each "note" off of your finger tips. Same process but much more complex. I noticed that once mankind created the "speaker" we have silver up any other way of ~~detecting~~ ~~transmitting~~ sound and speech.

Magnetic manipulation, will be the answer to this. and will take over Molassan the technology to the next and complete level.

IDea, 152 Permanent Hologram & Activated Hologram

We established true hologram technology, now we need to apply it to either be permanent or activated. This question is determined by infrastructure, goal of the hologram, and whether or not it is plugged into a direct power supply. This "IDea" would speak greatly to the consumer market, based on the products that would be made.

Idea #52 Permanent Hologram and activated Hologram.

refer to Idea, 149.

"Does Not Require Power"

A permanent Hologram is comprised from a SPM (special magnet) and its specific design and or features. For example, the colors being filtered, the "CAD CAM" design of your Hologram, and the percentage of ~~then~~ resistance. It may or may not have, whatever structure, pattern, or design you build into the SPM (special magnet) will not be able to take different shapes, designs, colors, resistance levels, or movements and or actions. It is a fixed creation that cannot be turned off. ~~an~~

A Natural activation for a permanent Hologram would ~~take~~ simply be ~~fish~~ light shining through the SPM (special magnet) field. Although the resistance and

motion ~~the~~ movement of the design.
                of action

of artificial

movements or actions would always remain even if you cannot "see" it. "Requires Power"

An Activated Holostatt would be able to undergo limitless updates to its designs, colors, actions or movements, and resistance-levels. ~~It would posess a sort of connectivity~~ An Activated Holostatt would <u>have</u> to be connected to some sort of _electrical_ power source due to the demands and or updates that would require more power than that of the Natural SPM (spacial magnets) power of its field _and_ or resistance.

Versitility requires power.

IDea, 153 & IDea, 154 & IDea, 155 & IDea, 157

For the next 4 "IDea's", I joined them together on the bases of possible consumer products. There is not much needed explanations for these "IDea's". What you see are examples of possible application. Imagine a world where these products are purchased at your local grocery store. Imagine how it would change your lifestyle if you had these products.

Idea, 153 — instead of A.I · Idea 149 (next to)
"Robotics" ... A.I
Holoframed humanoids /
"Hologram clones"

Idea, 154 video games and special effects & Applications "brought to life" via Hologram. refer to Idea, 149

The Magic of Disney (Mickey Mouse) brought to life via Hologram

Idea, 155 Hologram, permanent Idea, 149
and or activated "implants." // Idea, 152 refer to
From clothing to Facial augmentation

// = "Implants or wearable's"

Idea 157 — Hologram embedded smart "glove" for a "physical" smart phone.

LED lights and "special" magnets embedded into glove.

battery
charging port
power button / bluetooth connection.

IDea, 156

This next "IDea" is revolutionary but very high risk. Once consumer products are available and hologram implants are apart of the "norm". The question come up again whether the implants are permanent holograms or activated holograms. I think the consumer market would prefer, if not love, the flexibility of activated hologram implants. You can scroll on your phone to choose what ever you wanted to be a hologram. This feature does require a power source. You may have to sleep close to the phone charger at night to plug yourself in and charge yourself. However, there might be a controversial solution to that. Allow me to present "IDea, 156".

Idea, 156 *refer back to Idea, 149 *refer back to Idea, 152 Idea, 155.

implants on main arteries of the heart, to generate electricity.

Faster your heart pumps... the more electricity.

4 paddles per panel.   2 panels   "constant" blood flow

rotation

From the moment when our heart is created within us, our heart will pump or "beat" constantly throughout the course of our life. What if there was a way to harness that energy or power much like our hydroturbines? This technology would be ideal for giving electrical power to any type of implanted technology, such as "implanted" hologram technology.

IDea, 150

Since we recently got medical with the "IDea's" I am now jumping right into my last application for Magnetic Manipulation. This next "IDea" takes us closer to our health. The year 2020 has shown the world that viruses can come from anywhere at anytime. A lot of people have unfortunately died but many lives were saved with the help of masks and vaccines. In all seriousness, I felt that Magnetic Manipulation can help tremendously with our health. I present my final Magnetic Manipulation application.

Idea, 150 — Magnetic Manipulation to protect the cells in our body from viruses, pathogens, disease.

Through different applications of Magnetic Manipulation, we have figured out how to program certain commands or codes into "special macros". With "magnetic manipulation" we can "detect" everything with a nucleus, protons, and neutrons. This includes the ability to detect the cells in our body. With our cells detected, magnetic manipulation we can then (with the use of) create a magnetic shield or barrier around each cell in our body. With a certain code, of mission prestoned within the masses we can barricade and shield our cells from anything harmful that tries to feed off of our cell and spread throughout our body. With viruses unable to latch onto our cells and feed those viruses would starve out" theoretically. This would protect our cells from anything and everything that already exists or that would be completely brand new. Never have to come up with vaccines in a certain time frame while millions get infected and die. This innovation "is" mankinds future. Lets make it today.

1/ This technological revolution might take the shape of an implant to be located in our bones throughout our body. You would never need vaccines or hard when traveling to vast distances for a long amount of time.

"Through different applications of Magnetic Manipulation, we have figured out how to "program" certain commands or codes into "Special Magnets". With Magnetic Manipulation we can "detect" everything and anything with a nucleus, protons and neutrons. This includes the ability to detect the cells in our body. With our cells detected, with the use of Magnetic Manipulation, we can then create a magnetic "shield" around each cell in our body. With a certain code, or mission, programmed within the magnet, we can barricade and shield our cells from anything harmful that tries to feed off of our cells and spread throughout our body. With viruses unable to latch onto our cells, those viruses would speculatively "starve and die"

This concept would not only protect our cells from everything that already exists but also will protect our cells from anything completely brand new, regardless of the lethality the virus may possess. The world will "never" have to desperately seek out a vaccine while hundreds and thousands of people get infected and die.

This technological revolution might take the shape of an implant, or several implants, to be placed into our bones throughout our body."

Final Thoughts

For my final thoughts, I want to bring us back to Magnetic Manipulation, Magnetic Filtration and the considerable answer for space flight. Now that our minds are back on this topic, I want to explain exactly how we will "touch the stars". We will first use the Magnetic Manipulation to "repel" any and all universal influences. By doing this, we would theoretically create a tear or a "hole" in space and time. Once this "hole" is established, we would initiate the Magnetic Filtration. With all of the universal influences repelled, we would magnetically filter "in" light from a specific star.

To make since of what I am suggesting, let me further explain. The Magnetic Manipulation would create the wormhole and the Magnetic Filtration would give that wormhole a "heading" or a since of direction. Since light is a "constant", that means that the connection between us and any visible star is also a constant.

The wormhole, once activated, will accept the "constant" light along with the "constant connection", then "seek out" the source of that light. This in turn "teleports" point A to point B. To take it a step further, I speculate that depending on the percentage of filtered light from a specific star would dictate the certain distance away we are from the source of light. 100% filtered light might take us all the way to the surface of the star, which would be terrible for us. However, 80% filtered light might pull us back to a much safer distance away from the star to allow us to explore that solar system.

Of course, I am just a student with this new discovery. You see, to discover something brand new is to be the first and original student for new knowledge. I have learned as much as I can. However, in order to know if I am right or not or if my head is in the right direction, I need help.

I need my book to be read and analyzed by experts. I am also looking to you, my readers, to please share your thoughts on my book. Go on your social media's and leave a public review of my book. I truly hope I have inspired all of you with my honesty, humility, humbleness, and excitement with all of my "IDea's". I hope you all feel inspired to get your own ideas out there. Share with the world your voice and your mind. If you decide to publish your book on your ideas, maybe I can help you protect those ideas with my last "IDea"

IDea, 161 Protection Through Book Publishing

Idea, 161 "Protection through book publishing"

Before this book was made (many years ago), I had a challenge. To figure out a way to protect my ideas from being stolen by big corporations. My first thought, many years ago was through the Patenting office. After researching I came to realize it was incredibly expensive. I don't have enough money to set aside, not even for just "One" idea. The answer came to me early 2020. What if I used copy right law? Can copy right law offer me protection? Can copy right law "prevent" big corporations from stealing my ideas and patenting them themselves? I believe it can. Which led me through an affordable process. Book publishing through Amazon.

Amazon offers to publish your manuscript for free. As long as the book underlines explanation of the ideas, the intent of the ideas, and at few examples of applications for the ideas. But before you get that far you must prove that it is "you" who wrote or recorded, in some way the "New and original" ideas, that you are including in your book. My answer for this is through Notary from the UPS store. You schedule an appointment there and the can Notarize each idea, in document form or written form, for only $15 for each Notary. The Notary will record your name and other information and cannot it to the document of the idea. This will be your official prove that it is you who wrote the document.

Next, include the Notarized copy in your book. Taking these steps will assure no doubt for these ideas being yours. As I write this idea, my book is not published yet but I am hoping that these steps will protect the way that I think. if "successful" I want my method to be used by anyone and everyone with imagination, creativity, passion, and eagerness for voices to be heard. I want this to be the solution for the inequities of the patenting office. I believe everyone deserves to be recognized and given credit for any type of innovation, theories, designs, social changes, or paths towards a better life for all. All of this should "not" be exclusive for the rich but inclusive for everyone.

"Before this book was made, many years ago, I had a challenge. To figure out a way to protect my ideas from being stolen by big corporations. My first thought, many years ago, was through the patenting office. After researching I come to realize it was incredibly expensive. I don't have enough money to set aside, not even for just "one" idea. The answer came to me early 2020. What if I used copy right law? Can copy right law offer that protection? Can copy right law "prevent" big corporation from taking my ideas and patenting them themselves? I believe it can. Which led me through an affordable process. Book publishing through Amazon. Amazon offers to publish your manuscript for free. As long as the book underlines explanation of the ideas, the intent of the ideas, and a few examples of applications for the ideas. But before you get that far you must prove that it is "you" who wrote or recorded in some way the "new and original" ideas that you are including in your book. My answer for this through Notary form the UPS store. You schedule an appointment and they can Notarize each idea, in document form or written form, for only $15 for each Notary. The Notary will record your name and other information and connect it to the document of the idea. This will be your official proof that it is you who wrote the document. Next, include the Notarized copy in your book. Taking these steps will assure no doubt for those ideas being yours. As I write this "IDea", my book is not published yet. But I am hoping that these steps will protect in the way that I think. If successful, I want my method to be used by anyone and everyone with imagination, creativity, passion, and eagerness for voices to be heard. I want this to be the solution for the inequities of the patenting office. I believe everyone deserves to

be recognized and given credit for any type of innovation, theories, designs, social changes, or paths towards a better life for all. All of this should "Not" be exclusive for the rich but "inclusive" for everyone".

If you choose to go down this route of possible protection of your ideas.if you choose to use my method. Just make sure to put ,somewhere on the cover of your book, "protected by Z.S IDea, 161". This will respect the method I created and your use of that method. Thank you for your kindness.

I have included in my book a list of relevant people whom I would love to have that conversation with regarding my book. From computer engineers, to scientists, and to potential investors. Also on this list I have included a few reputable news outlets and journalists to help shine the light on my book and maybe help with the kinds of conversations I am looking for. The excitement I have for these "IDea's" are enough for me to go to such lengths. So please as a reader, share your thoughts. Tell your friends, your family, even your neighbors. Start your own conversations. Maybe use my book to amplify or inspire your own thoughts and innovations.

Science is an amazing thing and there is so much space to discover something brand new. If there was anything that I would suggest to my readers it would be to question everything. I do not care how simple or small it may appear, question it for maybe certain things have never been questioned before or maybe the right question has never been asked. And if you aren't given the answer you are looking for, or if the answer does not exist yet, then maybe you are on the path for your very own discovery of something brand new.

"Thank you for reading my book and I hope you walked away with more questions than you did coming into it". -Zachary Sanger

Relevant People Who I "<u>Invite</u>" to Read My Book

- My Family
- My Friends
- Neil deGrasse Tyson
- Michio Kaku
- Stephen Hawking
- James Cameron
- Morgan Freeman
- Will Smith
- Jeff Goldblum
- Keanu Reeves
- Michael J. Fox
- Stephen Colbert
- Jimmy Fallon
- NASA
- SpaceX
- D.A.R.P.A
- APPLE

- Microsoft
- Google
- Amazon
- Bungi
- M.I.T
- Berkely
- Harvard
- 60 Minutes
- N.P.R
- B.B.C
- C.N.N
- M.S.N.B.C
- The New York Times
- The Tonight Show
- Time Magazine
- Popular Science
- Popular Mechanics
- Science Channel
- History Channel's "Ancient Aliens"

"Potential Patents Pending in Progress Soon" for the following "IDea" numbers:

- 46
- 53
- 54
- 56
- 57
- 58
- 59
- 61
- 63
- 65
- 68
- 71
- 78
- 79
- 100
- 102

- 110
- 114
- 133
- 138
- 140
- 143
- 144
- 149
- 150
- 151
- 152
- 153
- 154
- 155
- 156
- 157
- 158
- 159
- 160

- 161

VERIFICATION AND PROOF OF AUTHENTICITY

9/1/2021

Zachary Thomas Hagstrom-Sanger
Sausalito, CA 94965

I Zachary Thomas Hagstrom-Sanger do solemnly swear that all of the contents within my book "Zachary Sanger's Special Theory of Magnetic Relativity" are original and without 3rd party involvement of any kind. I Zachary Thomas Hagstrom-Sanger wish to outline a few specific items from which come straight out of the book "Zachary Sanger's Special Theory of Magnetic Relativity". I have exhibits A, B and C for proof of my authenticity and originality.

Exhibit A, we have a black leather journal, hand made in Italy, with the code (A8188 CK34 CIAK BLACK) on a white sicker on the lower left corner on the back of the journal. Within Exhibit A, the first page states " Zach's Sketch book or drawings of ideas for Future" with three underlines below the word Future. Additionally, within Exhibit A, we have "IDea's" that range from "IDea, 1." all the way to "IDea, 106". A few specific items that I wish to highlight from Exhibit A, due to there relationship to the book " Zachary Sanger's Special Theory of Magnetic Relativity", will be listed by "IDea' numbers. Those "IDea" numbers are the following; 46, 53, 54, 56, 57, 58, 59, 61, 63, 68, 71, 78, 79, 100, and 102. Each of these "IDea" numbers represent a hand written and/or hand illustrated and/or explanation of my original thoughts and/or original ideas and/or theories.

Exhibit B, we have a brown leather journal, with the branded name on the front being "Comfy Strap" with the brands symbol above being a capital letter C and a capital letter S overlapping each other. The first page of Exhibit B contains the date June 17th, 2018 on the upper right corner of the first page. The first 25% of exhibit B are dated events. Then we have "IDea's" that range from "IDea, 107" all the way to "IDea, 147". A few specific items that I wish to highlight from Exhibit B, due to there relationship to the book "Zachary Sanger's Special Theory of Magnetic Relativity", will be listed by "IDea" numbers. Those "IDea" numbers are the following; 110, 114, 133, 138, 140, 143, and 144. Each of these "IDea" numbers represents a hand written and/or hand illustrated and/or explanation of my original thoughts and/or original ideas and/or original theories.

Exhibit C, we have a black leather journal, on the very last page of the journal has the STAMPED brand name. I will state what is exactly seen on that STAMP on the very last page, begging from top to bottom. First, we have an illustration of a hand holding a feather feather pen. Next, it states "made in Italy for Cavallini & Co. San Francisco". Lastly, a copy right symbol followed by "2003 Cavallini Papers & Co.Inc.". Within Exhibit C, we have "IDea" numbers ranging from "IDea, 148" all the way to "IDea, 161". A few specific items I wish to highlight from Exhibit C, due to there relationship to the book "Zachary Sanger's Special Theory of Magnetic Relativity", will be listed by "IDea" numbers. Those "IDea" numbers are the following; 149, 150, 151, 152, 153, 154, 155, 156, 157, 157, 158,159, 160, and 161. Each of these "IDea" numbers represents a hand written and/or hand illustrated and/or explanation of my original thoughts and/or original ideas and/or original theories.

I will additionally include my statements and explanations, directly from my book "Zachary Sanger's Special Theory of Magnetic Relativity", for which I refer to my "IDea'" on "Artificial Texture with hologram Via Magnetic Manipulation". This is necessary to include due to this specific "IDea" not coming out of one of my 3 journals, but came during the writing of the book itself. I promise authenticity and originality amongst all of my "IDea's" including this "IDea" as well.

My intent is clear and simple, to safeguard and insure protection for all of my "IDea's", explanations, and theories. My intent is to take this extra measure of protection to "PREVENT" infringement and to "PREVENT" research, development, sales and/or services without my involvement of any kind. Additionally, every "single" and "individual" "IDea" is available for lease. If sincerely interested in leasing out any "one" and/ or "multiple" "IDea's" to be used for research, development, sales and/or service please email me at zacharysanger@icloud.com. Serious inquiries only. When you email me you also are emailing my wife/manager Gayle Uyuni. Thank you for your respect.

To verify and authenticate my claims I have as my witness my wife Gayle Uyuni (Signature) _____ and an additional witness from the UPS Notary Department (Signature) _____

Date of Notary

Signature of Zachary Thomas Hagstrom-Sanger

CALIFORNIA ALL- PURPOSE
CERTIFICATE OF ACKNOWLEDGMENT

> A notary public or other officer completing this certificate verifies only the identity of the individual who signed the document to which this certificate is attached, and not the truthfulness, accuracy, or validity of that document.

State of California }

County of Marin }

On 09·01·2021 before me, Evan Timmel, Notary Public
(Here insert name and title of the officer)

personally appeared Zachary Thomas Hagstrom Sanger and Gayle Uyuni,
who proved to me on the basis of satisfactory evidence to be the person(s) whose name(s) is/are subscribed to the within instrument and acknowledged to me that he/she/they executed the same in his/her/their authorized capacity(ies), and that by his/her/their signature(s) on the instrument the person(s), or the entity upon behalf of which the person(s) acted, executed the instrument.

I certify under PENALTY OF PERJURY under the laws of the State of California that the foregoing paragraph is true and correct.

WITNESS my hand and official seal.

Evan Timmel
Notary Public Signature

(Notary Public Seal)
EVAN TIMMEL
Notary Public - California
Marin County
Commission # 2343341
My Comm. Expires Jan 25, 2025

ADDITIONAL OPTIONAL INFORMATION

DESCRIPTION OF THE ATTACHED DOCUMENT

Verification and Proof
(Title or description of attached document)

of Authenticity
(Title or description of attached document continued)

Number of Pages _____ Document Date _____

CAPACITY CLAIMED BY THE SIGNER
- ☐ Individual (s)
- ☐ Corporate Officer
  _____ (Title)
- ☐ Partner(s)
- ☐ Attorney-in-Fact
- ☐ Trustee(s)
- ☐ Other _____

2015 Version www.NotaryClasses.com 800-873-9865

INSTRUCTIONS FOR COMPLETING THIS FORM
This form complies with current California statutes regarding notary wording and, if needed, should be completed and attached to the document. Acknowledgments from other states may be completed for documents being sent to that state so long as the wording does not require the California notary to violate California notary law.

- State and County information must be the State and County where the document signer(s) personally appeared before the notary public for acknowledgment.
- Date of notarization must be the date that the signer(s) personally appeared which must also be the same date the acknowledgment is completed.
- The notary public must print his or her name as it appears within his or her commission followed by a comma and then your title (notary public).
- Print the name(s) of document signer(s) who personally appear at the time of notarization.
- Indicate the correct singular or plural forms by crossing off incorrect forms (i.e. he/she/they,- is /are) or circling the correct forms. Failure to correctly indicate this information may lead to rejection of document recording.
- The notary seal impression must be clear and photographically reproducible. Impression must not cover text or lines. If seal impression smudges, re-seal if a sufficient area permits, otherwise complete a different acknowledgment form.
- Signature of the notary public must match the signature on file with the office of the county clerk.
  ❖ Additional information is not required but could help to ensure this acknowledgment is not misused or attached to a different document.
  ❖ Indicate title or type of attached document, number of pages and date.
  ❖ Indicate the capacity claimed by the signer. If the claimed capacity is a corporate officer, indicate the title (i.e. CEO, CFO, Secretary).
- Securely attach this document to the signed document with a staple.

www.ingramcontent.com/pod-product-compliance
Lightning Source LLC
Chambersburg PA
CBHW082109220526
45472CB00009B/2110